Springer Series in
Nuclear
and **Particle Physics**

Springer Series in **Nuclear** and **Particle Physics**

Editors: Mary K. Gaillard · J. Maxwell Irvine · Erich Lohrmann · Vera Lüth
Achim Richter

Hasse, R. W., Myers W. D.
Geometrical Relationships of Macroscopic Nuclear Physics

Belyaev, V. B.
Lectures on the Theory of Few-Body Systems

Heyde, K.
The Nuclear Shell Model

Gitman, D. M., Tyutin I. V.
Quantization of Fields with Constraints

Sitenko, A. G.
Scattering Theory

V. B. Belyaev

Lectures on the Theory of
Few-Body Systems

With 31 Figures

Springer-Verlag Berlin Heidelberg New York
London Paris Tokyo Hong Kong

Professor Dr. Vladimir B. Belyaev
Laboratorium of Theoretical Physics,
JINR 141980 Dubna, USSR

Translator:
Dr. G. B. Pontecorvo
Laboratorium of Nuclear Problems,
JINR 141980 Dubna, USSR

Title of the original Russian edition: Lektsii po teorii malochastichnuikh sistem
© Energoatomizdat, Moskow 1986

ISBN 978-3-642-87294-5   ISBN 978-3-642-87292-1 (eBook)
DOI 10.1007/978-3-642-87292-1

Foreword

Nuclear physics is undoubtedly a many-body problem. A nice introduction into the present status of this subject may be found in the comprehensive monograph by P. Ring and P. Schuck "The Nuclear Many-Body Problem" (Springer, Berlin, Heidelberg, New York 1980). However, in view of the many challenging problems that remain to be tackled, it is sensible to consider systems with few particles as model cases. These provide the basis for solving the sophisticated many-body problem posed by intermediate and heavy nuclei.

Out of the large number of existing nuclear systems, few-particle, that is few-nucleon, systems can be singled out to form a special group. This is possible because a comparatively small number of degrees of freedom (or dynamic variables) is required for a complete description of such systems.

In these Lectures we utilize this to study few-body systems in great detail, in particular three- and four-body systems. In contrast to published monographs on the subject, we deal not just with nucleonic degrees of freedom but consider also non-nucleonic degrees of freedom. The range of approaches and methods examined exceeds the scope of other textbooks.

The Lectures are organized in such a way as to guide the uninitiated reader through the essentials of solving the dynamical equations of few-body systems directly towards practical applications. Formally oriented readers might like to supplement their reading with texts such as "The Quantum Mechanical Few-Body Problem" by W. Glöckle (Springer, Berlin, Heidelberg, New York 1983). However, the basis provided by these Lectures is sufficiently broad to facilitate independent work in real problems without consulting further literature.

Dubna, February 1990   *V.B. Belyaev*

Preface

In three- and four-body problems it is possible to obtain mathematically correct and computationally tractable equations such as the Faddeev and Yakubovsky equations describing exactly, for any assumed interaction between the particles, the motion of few-nucleon systems. Their physical characteristics are derived without the introduction of any a priori ideas of the motion of particles, solely on the basis of "first" principles, however, with the assumption that only nucleonic degrees of freedom are important; non-nucleonic degrees of freedom are discussed separately. A significant part of these lectures is therefore devoted to an exposition of the mathematics of the Faddeev and Yakubovsky equations and of their modifications. Various approximate methods used for the practical solution of these equations are also discussed. While the three- and four-body problems are formulated rigorously, these lectures also present examples of approximate dynamic equations whose solution does not require the sophisticated techniques necessary for solving the exact equations.

Throughout practically the whole of physics, an understanding of the properties of a physical system is achieved when one succeeds in representing the system as a one-particle system. Examples of such reductions are the Hartree-Fock equations and the method of quasi-particles, or the Chew-Low equations in the field-theory description of the πN-interaction. Attempts to achieve such a reduction, however, are not always successful, since systems with three or more particles may possess qualitative peculiarities going beyond the simple one-particle picture. Three examples of such properties are

1. The so-called Efimov effect arising in a system of three particles interacting through a short-range potential (for instance, a square-well potential), such that in a pair subsystem there exists a level of zero or close to zero energy. In this case a situation can arise in which the levels of the system are "pushed" out of the well as the depth of the potential increases, in contrast to the situation in a two-body problem.

2. The collapse occurring in a system of three particles interacting through a pair δ-like potential (Thomas effect). Nothing of the sort happens in a two-particle system.

3. Off-mass-shell characteristics of the pair T-matrix, indicating that for a complete description of a system consisting of three or more particles it is insufficient to know only the two-particle scattering phases or, which is the same thing, the on-shell T-matrix. Here we see the requirement of utilizing new characteristics of pair interaction that are not observable in the two-body problem.

The material of these lectures is arranged so that the reader can obtain information needed for computation of the characteristics of quantum-mechanical systems with a small number of degrees of freedom. Therefore numerous formulas are encountered. The author has dealt in detail with approximate methods of solution of the two-body problem. This is done not only for the subsequent application of these approximations to the three-body problem, but also to update the 40-to-50-year-old expositions of these methods (perturbation theory, Born approximation and others) normally presented in courses on quantum mechanics.

The material generally follows the lectures given by the author over a period of several years to students of theoretical physics at the Moscow State University and at other universities. A significant part is based on results scattered in numerous original papers, so that the content of the lectures has little in common with those few textbooks that have already been published. These are listed under General Reading.

The author is grateful in advance to those readers who will find it worthwhile to express comments concerning the content of the lectures.

Dubna, February 1990   *V.B.Belyaev*

Contents

1. The Two-Body Problem

In this lecture, the equation of motion (the Lippman–Schwinger equation) is presented and relations are given for the pair t-matrix, which are satisfied for a wide class of interaction potentials between colliding particles. A significant part is devoted to approximate methods of solution of the two-body problem, based on finite-dimensional approximation (separable expansions) of the interaction potential.

1.1 Properties of the Two-Particle t-Matrix

In this lecture we shall deal only with two-particle systems. Ultimately, we shall be interested in the characteristics of the scattering problem, such as the scattering amplitude and the off-shell t-matrix, as well as the spectrum of bound states. In the last part of the lecture it is shown that to find the spectrum of eigenvalues of a system there is absolutely no need to match the function within the range of the potential with the external function, as is usually recommended in textbooks. In the same way the scattering length of two particles interacting, for example, by means of the Yukawa potential $\exp(-\mu r)/r$, can be found without using perturbation theory in the potential, nor any other standard approximation.

Let us start with the Schrödinger equation for two particles in the centre-of-mass system,

$$(H_0 + V - E)\psi(\boldsymbol{r}) = 0 , \tag{1.1}$$

where H_0 is the kinetic energy operator of the relative motion of the particles; V is the interaction potential (operator)

$$V\psi = \int \langle \boldsymbol{r}|V|\boldsymbol{r}'\rangle \psi(\boldsymbol{r}')d\boldsymbol{r}' . \tag{1.2}$$

Most of the calculations below are carried out in the momentum representation. Therefore, we introduce the Fourier transform of the pair potential,

$$\langle \boldsymbol{k}|V|\boldsymbol{k}'\rangle = \int \exp(-i\boldsymbol{k}\boldsymbol{r})\langle \boldsymbol{r}|V|\boldsymbol{r}'\rangle \exp(i\boldsymbol{k}'\boldsymbol{r}')d\boldsymbol{r}\ d\boldsymbol{r}' . \tag{1.3}$$

For a local potential

$$\langle \boldsymbol{r}|V|\boldsymbol{r}'\rangle = V(r)\delta(\boldsymbol{r} - \boldsymbol{r}') \tag{1.4}$$

and, obviously, the Fourier transform in this case takes the form

$$\langle \mathbf{k}|V|\mathbf{k}'\rangle = \int V(r)\exp[i(\mathbf{k}-\mathbf{k}')r]d\mathbf{r} \ . \tag{1.5}$$

Expression (1.5) is convenient for obtaining the partial harmonics of the Fourier transform of the potential. Indeed, the l-th harmonic is found simply by expanding the exponentials in the right-hand part into spherical functions and integration over the angles.

Instead of the Schrödinger equation (1.1), we shall make use of two integral equations, taking into account the boundary conditions. In the case of bound states, the wave function is

$$\psi(\mathbf{k}, E) = -\frac{2\mu}{k^2+\alpha^2}\int \langle \mathbf{k}|V|\mathbf{k}'\rangle\psi(\mathbf{k}', E)\frac{d\mathbf{k}'}{(2\pi)^3} \ , \tag{1.6}$$

where $E = -\alpha^2/(2\mu) < 0$ is the energy of the system; μ is the reduced mass of the two particles.

For the scattering problem the equation has the form

$$\psi_{\mathbf{k}_0}(\mathbf{k}, Z) = (2\pi)^3\delta(\mathbf{k}-\mathbf{k}_0)$$
$$+ \left(E - \frac{k^2}{2\mu} + i\varepsilon\right)^{-1}\int \langle \mathbf{k}|V|\mathbf{k}'\rangle\psi_{\mathbf{k}_0}(\mathbf{k}', Z)d\mathbf{k}'/(2\pi)^3 \ , \tag{1.7}$$

where $E = k_0^2/2\mu > 0$; $Z = E + i\varepsilon$.

Note that in the scattering wave function $\psi_{\mathbf{k}_0}(\mathbf{k}, Z)$ the vector $\mathbf{k}$ represents the dynamic variable, while the vector $\mathbf{k}_0$ is a parameter describing the momentum associated with the incident plane wave. We now introduce the pair t-matrix by the relation

$$\langle \mathbf{k}|t(Z)|\mathbf{k}_0\rangle \equiv \int \langle \mathbf{k}|V|\mathbf{k}'\rangle\psi_{\mathbf{k}_0}(\mathbf{k}', Z)d\mathbf{k}'/(2\pi)^3$$
$$= \int \exp(-i\mathbf{k}r)V(r)\psi_{\mathbf{k}_0}(r, Z)d\mathbf{r} \ . \tag{1.8}$$

An important consequence of the second integral in (1.8) is, because we shall always deal with short-range potentials, that for calculation of the pair t-matrix it is only necessary to find the scattering wave function $\psi_{\mathbf{k}_0}(r, Z)$ within the range of the potential.

Using the relation (1.8) in (1.7) we obtain the following useful relationship between the scattering wave function and the t-matrix:

$$\psi_{\mathbf{k}_0}(\mathbf{k}, Z) = (2\pi)^3\delta(\mathbf{k}-\mathbf{k}_0) + \frac{\langle \mathbf{k}|t(Z)|\mathbf{k}_0\rangle}{E - k^2/(2\mu) + i\varepsilon} \ . \tag{1.9}$$

From the definition of the pair t-matrix (1.8) and the equation for the wave function (1.7) we readily obtain the well-known *Lippmann–Schwinger equation for the pair t-matrix*:

$$\langle k|t(Z)|k'\rangle = \langle k|V|k'\rangle + \int \frac{\langle k|V|q\rangle}{E - q^2/2(\mu) + i\varepsilon}\langle q|t(Z)|k\rangle\frac{dq}{(2\pi)^3} . \qquad (1.10)$$

Here, generally speaking, $k^2 \neq k'^2 \neq 2\mu E$.

The scattering amplitude f is related to the t-matrix by

$$f(k, k') = -(\mu/2\pi)\langle k|t(E)|k'\rangle \qquad (1.11)$$

for $E = k^2/(2\mu) = k'^2/(2\mu)$.

We shall call the t-matrix in the right-hand side of (1.11) the *on-mass-shell t-matrix*, which is distinguished from the solution of (1.10), the *off-shell t-matrix*.

If the interaction V is central, then the Hamiltonian of the system is known to be invariant with respect to rotations, i.e. angular momentum is conserved in the system. As a result, instead of (1.10), we obtain the following equation for the partial amplitude $t_l(k, k', Z)$ with the inhomogeneous term $V_l(k, k')$:

$$V_l(k, k') = 4\pi \int j_l(kr)V(r)j_l(k'r)r^2 dr , \qquad (1.12)$$

where $j_l(x)$ is the spherical Bessel function

$$j_l(x) = \sqrt{\pi/(2x)}J_{l+1/2}(x) .$$

The partial off-shell t-matrix $t_l(k, k', Z)$ satisfies the generalized unitarity relation

$$\text{Im}\, t_l(k, k', Z) = -\frac{\mu q}{2\pi}t_l(k, q, Z)t_l^*(k', q, Z) , \qquad (1.13)$$

where $q^2/2\mu = E$.

Equation (1.13) transforms on the mass shell into the usual unitarity relation for the two-particle scattering amplitude. This relation can be obtained from the so-called *Hilbert identity for the t-matrix*, which we shall now derive.

Let the Hamiltonian of the system be

$$H = H_0 + V . \qquad (1.14)$$

We introduce the Green functions of the system $G(Z)$ and $G_0(Z)$:

$$G(Z) = (Z - H)^{-1} ; \quad G_0(Z) = (Z - H_0)^{-1} . \qquad (1.15)$$

From (1.15) we readily verify that the Green function of the system $G(Z)$ satisfies the operator equation (or "second resolvent equation")

$$G(Z) = G_0(Z) + G_0(Z)VG(Z) . \qquad (1.16)$$

One can introduce an alternative, with respect to (1.8), definition of the pair t-matrix:

$$t(Z) = V + VG(Z)V . \qquad (1.17)$$

Indeed, by using (1.16, 17) we arrive at the Lippmann–Schwinger equation for the t-matrix in operator form:

$$t(Z) = V + VG_0(Z)t(Z) \ . \tag{1.18}$$

From (1.17, 18) we easily obtain the reciprocal relation between the Green function and the t-matrix:

$$G(Z) = G_0(Z) + G_0(Z)t(Z)G_0(Z) \ . \tag{1.19}$$

From (1.19), one can see that in addition to those of the t-matrix the Green function exhibits additional kinematic singularities in the energy Z. As follows from (1.19), they are related to the singularities of the free Green function $G_0(Z)$.

It is easy to check that the following Hilbert identity (or "first resolvent equation") holds at energies Z_1 and Z_2:

$$G(Z_1) - G(Z_2) = (Z_1 - Z_2)G(Z_1)G(Z_2) \ . \tag{1.20}$$

(To verify, multiply (1.20) on the left by $G^{-1}(Z_1)$ and on the right by $G^{-1}(Z_2)$.) Let us obtain the analogous identity for the t-matrix. To this end we shall show that the following relations hold:

$$t(Z)G_0(Z) = VG(Z) \tag{1.21}$$

and

$$G_0(Z)t(Z) = G(Z)V \ . \tag{1.22}$$

We now multiply (1.19) on the left by V:

$$VG(Z) = VG_0(Z) + VG_0(Z)t(Z)G_0(Z) = [V + VG_0(Z)t(Z)]G_0(Z)$$
$$= t(Z)G_0(Z) \ ,$$

and thus we obtain (1.21). Further, we multiply (1.17) on the left by $G_0(Z)$:

$$G_0(Z)t(Z) = G_0(Z)V + G_0(Z)VG(Z)V$$
$$= [G_0(Z) + G_0(Z)VG(Z)]V = G(Z)V \ ,$$

i.e., we have (1.22).

Now, multiplying the Hilbert identity (1.20) on the left and the right by V we obtain

$$t(Z_1) - t(Z_2) = (Z_2 - Z_1)VG(Z_1)G(Z_2)V$$
$$= (Z_2 - Z_1)t(Z_1)G_0(Z_1)G_0(Z_2)t(Z_2) \ . \tag{1.23}$$

Because they are operator equations, all the formulae from (1.14) to (1.23) hold not only for two-particle systems but for systems with arbitrary numbers of particles.

Transforming from the operator to the momentum representation we arrive at the *Hilbert identity for t-matrices* in its final form:

$$t(\boldsymbol{k}, \boldsymbol{k}', Z_1) - t(\boldsymbol{k}, \boldsymbol{k}', Z_2)$$
$$= (Z_2 - Z_1) \int \frac{d\boldsymbol{q}}{(2\pi)^3} t(\boldsymbol{k}, \boldsymbol{q}, Z_1) \frac{t(\boldsymbol{q}, \boldsymbol{k}', Z_2)}{[q^2/(2\mu) - Z_1][q^2 1(2\mu) - Z_2]} \ . \tag{1.24}$$

We now demonstrate that in a system consisting of two particles capable of forming a single bound state, the t-matrix is [1]

$$t(\boldsymbol{k}, \boldsymbol{k}', Z) = \frac{\varphi(\boldsymbol{k})\varphi(\boldsymbol{k}')}{Z + \kappa^2} + V(\boldsymbol{k} - \boldsymbol{k}') + \int \frac{d\boldsymbol{q}}{(2\pi)^3}$$
$$\times t\left(\boldsymbol{k}, \boldsymbol{q}\frac{q^2}{2\mu} \pm i0\right) \frac{1}{q^2/(2\mu) - Z}$$
$$\times t \cdot (\boldsymbol{q}, \boldsymbol{k}', q^2/2\mu \pm i0)^* .$$

(1.25)

Here $\kappa^2 = |E_0|$ is the energy of the bound state.

Since $t(Z)$ is an analytic function in the complex Z-plane, the following Cauchy formula holds:

$$t(\boldsymbol{k}, \boldsymbol{k}', Z) = \frac{1}{2\pi i} \int_\gamma \frac{t(\boldsymbol{k}, \boldsymbol{k}', S)}{S - Z} dS .$$

(1.26)

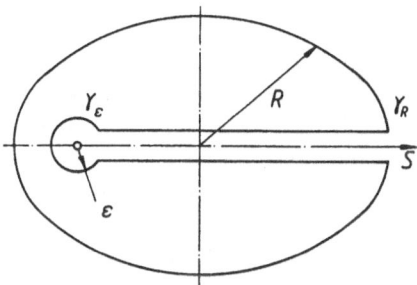

Fig. 1.1. Contour of integration γ in (1.26)

For the integral along the contour γ_ε (Fig. 1.1) we have

$$\lim_{\varepsilon \to 0} \frac{1}{2\pi i} \int_{\gamma_\varepsilon} \frac{t(\boldsymbol{k}, \boldsymbol{k}', S)}{S - Z} dS = \frac{\varphi(\boldsymbol{k})\varphi(\boldsymbol{k}')}{Z + \kappa^2} ,$$

(1.27)

where the function $\varphi(\boldsymbol{k})$, sometimes called the vertex form factor, is proportional to the wave function of the bound state. [Expression (1.27) may be obtained by utilizing the spectral expansion of the Green function and (1.17).] We shall use the estimate obtained in [1] which permits us to find the integral along the contour γ_R:

$$|t(\boldsymbol{k}, \boldsymbol{k}', Z) - V(|\boldsymbol{k} - \boldsymbol{k}'|)| \leq \frac{c}{[1 + |\boldsymbol{k} - \boldsymbol{k}'|]^{1+\theta}[1 + |z|]^{\theta/2}} ,$$

(1.28)

where θ is independent of Z.

Obviously, as $R \to \infty$, the integral along the contour γ_R just equals $V(|\boldsymbol{k} - \boldsymbol{k}'|)$. Finally, the last integral along the contour enclosing the cut involves the difference

$$t(\boldsymbol{k}, \boldsymbol{k}', S + \mathrm{i}0) - t(\boldsymbol{k}, \boldsymbol{k}', S - \mathrm{i}0) \ .$$

Let us now transform it, by using (1.24) and assuming therein

$$Z_1 = S + \mathrm{i}\varepsilon, Z_2 = S - \mathrm{i}\varepsilon; t(\boldsymbol{k}, \boldsymbol{k}', S + \mathrm{i}0) - t(\boldsymbol{k}, \boldsymbol{k}', S - \mathrm{i}0)$$

$$= \lim_{\varepsilon \to 0} \int t(\boldsymbol{k}, \boldsymbol{q}, S + \mathrm{i}\varepsilon) \frac{2\mathrm{i}\varepsilon}{[q^2/(2\mu) - S] + \varepsilon^2} t^*(\boldsymbol{q}, \boldsymbol{k}', S - \mathrm{i}\varepsilon) \frac{d\boldsymbol{q}}{(2\pi)^3}$$

$$= 2\pi\mathrm{i} \int t\left(\boldsymbol{k}, \boldsymbol{q}, \frac{q^2}{2\mu} \pm \mathrm{i}0\right) \times \delta\left(\frac{q^2}{2\mu} - S\right)$$

$$\times t^*\left(\boldsymbol{q}, \boldsymbol{k}', \frac{q^2}{2\mu} \mp \mathrm{i}0\right) \frac{d\boldsymbol{q}}{(2\pi)^3} \ , \tag{1.29}$$

where $\delta[q^2/(2\mu) - S]$ is the Dirac delta function.

Applying (1.29) in (1.26) we immediately obtain the third term in (1.25). Relations (1.24, 25) impose restrictions on the off-shell t-matrix.

We now derive one more representation for the t-matrix, the so-called *Noyes–Kowalski representation* [2]. In this representation the t-matrix is expressed as a sum of two terms, the first of which is separable and the second vanishes on the half-mass-shell surface. As a result, the t-matrix satisfying an equation involving a singular kernel (the Lippmann–Schwinger equation) is expressed in terms of two functions, each of which satisfies an equation with a non-singular kernel. Besides this, as shown in the second lecture, the Noyes–Kowalski representation permits separation of the three-particle wave function into an internal and an external part, without introduction of a characteristic range separating the space into an internal and an external part.

Thus, we write the partial t-matrix, for instance, its S wave component $t(k, k', Z)$, in the form

$$t(k, k', E + \mathrm{i}0) = \frac{t(k, q, E + \mathrm{i}0)t(q, k', E + \mathrm{i}0)}{t(q, q, Z + \mathrm{i}0)} + R(k, k', Z) \ , \tag{1.30}$$

where $Z = q^2/2\mu + \mathrm{i}0$, provided the denominator is different from zero. In the case of a potential V containing repulsive and attractive parts, the on-shell t-matrix can be equal to zero for some values of the energy. Reference [3] explains how (1.30, 32) must be treated in such a situation.

We introduce the notation

$$f(k, q) \equiv \frac{t(k, q, E + \mathrm{i}0)}{t(q, q, E + \mathrm{i}0)}; t(q) \equiv \left(q, q, \frac{q^2}{2\mu} + \mathrm{i}0\right) \ . \tag{1.31}$$

Obviously (1.30) can be written

$$t(k, k', E + \mathrm{i}0) = f(k, q)t(q)f(k', q) + R(k, k', Z) \ . \tag{1.32}$$

Let us obtain the equation for the function $f(k, q)$. From the Lippmann–Schwinger equation

$$t(k, k', Z) = V(k, k') + \frac{1}{2\pi^2} \int\limits_0^\infty V(k, p) \frac{t(p, k', Z)}{Z - p^2/(2\mu)} p^2 \, dp \,, \tag{1.33}$$

in which we set $k' = q$, we subtract the Lippmann–Schwinger equation in which $k = k' = q$ is multiplied by $V(k, q)/V(q, q)$ and obtain

$$f(k, q) = V(k, q)/V(q, q) + \int\limits_0^\infty \Lambda(k, p, q) \times \frac{1}{Z - p^2} f(p, q) \frac{p^2 dp}{2\pi^2} \,, \tag{1.34}$$

where $\Lambda(k, k', q) = V(k, k') - V(k, q)V(q, k')/V(q, q)$. Since, by definition, $\Lambda(k, q, q) = \Lambda(q, k', q) = 0$, the function $f(k, q)$ satisfies an equation with a non-singular kernel. Using (1.32–34) we also obtain an equation with a regular kernel for the function R:

$$R(k, k', Z) = \Lambda(k, k', q) + \int\limits_0^\infty \Lambda(k, p, q) \frac{R(p, k', Z)}{Z - p^2/(2\mu)} \frac{p^2 dp}{2\pi^2} \,. \tag{1.35}$$

From this equation we see that on the half-mass-shell, e.g., when $k = q$, $R = 0$. Thus, $R(k, k', Z)$ can be written

$$R(k, k', Z) = [k^2/(2\mu) - Z] r(k, k', Z) \,, \tag{1.36}$$

where the function $r(k, k', Z)$ no longer tends to zero at finite k. If $f(p, q)$ is found, then the scattering amplitude $t(q)$ is expressed through it in the form of a simple quadrature:

$$t(q) = V(q, q) \left[1 - \int\limits_0^\infty V(q, p) \frac{1}{E - p^2/2(\mu) + \mathrm{i}0} f(p, q) \frac{p^2 dp}{2\pi^2} \right]^{-1} \,.$$

This expression is extremely useful for finding the scattering amplitude with the aid of various approximate methods of computing $f(p, q)$. For example, already the zeroth iteration of the equation for function f, i.e., $V(p, q)/V(q, q)$, often leads to a quite reasonable approximation for the amplitude.

1.2 Phase-Equivalent Potentials

We shall now proceed to a more detailed discussion of the properties of the off-shell pair t-matrix. For this purpose we have to introduce several new concepts, such as the Möller wave operators, off-shell wave functions and phase-equivalent potentials. By applying these concepts, we shall first demonstrate that the off-shell continuation of the Schrödinger equation is ambiguous.

Thus, we start from the Schrödinger equation

$$(k^2 - H)|\psi_k\rangle = 0 . \tag{1.37}$$

Here again $k^2 = E; H = H_0 + V$.

We shall assume that two conditions are fulfilled:

a) the decrease of the potential $V(r)$ is sufficiently rapid, and that it contains a finite number of bound states;

b) the boundary condition for $k^2 > 0$

$$\langle r|\psi_k\rangle \equiv \psi_k(r)_{r \to \infty} = \langle k|r\rangle + t \exp(ikr)/r . \tag{1.38}$$

As already pointed out, (1.37, 38) are equivalent to the Lippmann-Schwinger equation for the wave function

$$|\psi_{k,k^2}^{(+)}\rangle = |k\rangle + (k^2 - H_0 + i0)^{-1} V |\psi_{k,k^2}^{(+)}\rangle , \tag{1.39}$$

where $|k\rangle$ satisfies

$$(k^2 - H_0)|k\rangle = 0 . \tag{1.40}$$

We introduce the *Möller wave operator* $\Omega(Z)$ as the solution of the equation

$$\Omega(Z) = 1 + (Z - H_0)^{-1} V \Omega(Z) . \tag{1.41}$$

Obviously, the operator $\Omega(Z)$ can be represented as

$$\Omega(Z) = 1 + (Z - H)^{-1} V . \tag{1.42}$$

It now becomes clear that the Lippmann-Schwinger equation (1.39) may be written in the following form:

$$|\psi_{k,k^2}^{(+)}\rangle = \Omega(k + i0)|k\rangle . \tag{1.43}$$

The off-shell continuation of (1.43) is then

$$|\psi_{k',k^2}^{(+)}\rangle = \Omega(k^2 + i0)|k'\rangle , \quad k'^2 \neq k^2 . \tag{1.44}$$

The function $|\psi\rangle$ determined by (1.44) is called the *off-shell wave function*. It is no longer an eigenfunction of the Hamiltonian H, but satisfies

$$(k^2 - H)|\psi_{k',k^2}^{(+)}\rangle = (k^2 - k'^2)|k'\rangle . \tag{1.45}$$

We shall now show that by starting from the same Schrödinger equation (1.37), one can obtain the off-shell continuation of this equation in an infinite number of ways, i.e., it is not unique.

Let us rewrite (1.37) identically:

$$[k^2 - U^+ H_0 U - (H - U^+ H_0 U)]|\psi_k\rangle = 0 , \tag{1.46}$$

where U is an unitary operator approximating the unit operator in such a way that the vector $(U^+ - 1)|k\rangle$ is finite. Then, in the r-representation we have the condition

$$\lim_{r,r'\to\infty} rr'\langle r|U^+ - 1|r'\rangle \to 0 , \tag{1.47}$$

i.e., the operator U^+ is "localized" in the region of small r. We now obtain the off-shell continuation of (1.46) with the aid of the complete set of states $|\bar{k}\rangle$, where $|\bar{k}\rangle$ satisfies the equation

$$(k^2 - U^+ H_0 U)|\bar{k}\rangle = 0 . \tag{1.48}$$

From (1.40, 48) it follows that

$$|\bar{k}\rangle = U^+|k\rangle . \tag{1.49}$$

By analogy to (1.41), we introduce a new wave operator $\bar{\Omega}(Z)$ satisfying the operator equation

$$\bar{\Omega}(Z) = 1 + (Z - U^+ H_0 U)^{-1}(H - U^+ H_0 U)\bar{\Omega}(Z) \tag{1.50}$$

or

$$\bar{\Omega}(Z) = 1 + (Z - H)^{-1}(H - U^+ H_0 U) . \tag{1.51}$$

Thus, the off-shell continuation of (1.46) is obtained by

$$|\bar{\psi}^{(+)}_{k',k^2}\rangle = \bar{\Omega}(k^2 + i0)|\bar{k}'\rangle = \bar{\Omega}(k^2 + i0)U^+|k'\rangle . \tag{1.52}$$

From (1.44, 52) we get

$$|\bar{\psi}^{(+)}_{k',k^2}\rangle - |\psi^{(+)}_{k',k^2}\rangle = \bar{\Omega}(k^2 + i0)U^+ - \Omega(k^2 + i0)^\dagger]|k'\rangle . \tag{1.53}$$

Substituting the expressions for the wave operators into (1.53), we ultimately obtain

$$|\bar{\psi}^{(+)}_{k',k^2}\rangle - |\psi^{(+)}_{k',k^2}\rangle = (k^2 - k'^2)(k^2 - H + i0)^{-1}(U^+ - 1)|k'\rangle , \tag{1.54}$$

i.e., the continuation of the wave function to the region $k^2 \neq k'^2$ is unambiguous if, and only if, $U^+ = 1$.

All the observables in the two-body problem are determined on the mass shell, i.e., in the region of $k^2 = k'^2$. Therefore, any two continuations lead to identical observations.

We now introduce the concept of *phase-equivalent potentials*.

Referring to the initial Schrödinger equation (1.37) one can define the interaction as $V = H - H_0$ and the t-matrix in the form

$$\langle k''|T(k^2)|k'\rangle = \langle k''|V|\psi^{(+)}_{k',k^2}\rangle = \langle k''|V\Omega(k^2 + i0)|k'\rangle . \tag{1.55}$$

In (1.55) the t-matrix is determined in the basis of eigenfunctions $|k\rangle$ of the free Hamiltonian H_0. Starting from (1.46) we introduce, analogously, the "interaction" $\bar{V} = H - U^+ H_0 U$ and the "free" Hamiltonian $\bar{H}_0 = U^+ H_0 U$, the eigenfunctions of which, as assumed above, are the vectors $|\bar{k}\rangle$. We now introduce the t-matrix $\bar{T}$ corresponding to the interaction V:

$$\langle \bar{k}''|\bar{T}(k^2)\bar{k}'\rangle = \langle \bar{k}''|\bar{V}|\bar{\psi}^{(+)}_{k\prime,k^2}\rangle = \langle \bar{k}''|\bar{V}\bar{\Omega}(k^2+\mathrm{i}0)|\bar{k}'\rangle \ . \tag{1.56}$$

To compare the t-matrices (1.55, 56) we introduce the operator

$$\hat{T}(k^2) \equiv U\bar{T}(k^2)U^+ = U\bar{V}\bar{\Omega}(k^2+\mathrm{i}0)U^+ \ , \tag{1.57}$$

the matrix elements of which are determined in the basis $|k\rangle$. This operator will be of use to us further on. We also introduce the corresponding interaction operator $\hat{V}$ and the wave operator $\hat{\Omega}$:

$$\hat{V} = U\bar{V}U^+ = UHU^+ - H_0 \ ; \tag{1.58}$$

$$\hat{\Omega}(Z) = U\bar{\Omega}(Z)U^+ = 1 + (Z-H_0)^{-1}\hat{V}\hat{\Omega}(Z) \ . \tag{1.59}$$

Then (1.57) becomes

$$\hat{T}(k^2) = \hat{V}\hat{\Omega}(k^2+\mathrm{i}0) \ . \tag{1.60}$$

We can now show that all potentials determined by (1.58) exhibit identical properties on the mass shell. Indeed, under the unitarity transformation U, the initial Schrödinger equation (1.37) assumes the form

$$(k^2 - UHU^+)|\hat{\psi}^{(+)}_{k\prime,k^2}\rangle = 0 \ , \tag{1.61}$$

where

$$|\hat{\psi}^{(+)}_{k\prime,k^2}\rangle = U|\psi^{(+)}_{k\prime,k^2}\rangle \ . \tag{1.62}$$

Now we rewrite (1.61) by using (1.58):

$$(k^2 - H_0 - \hat{V})|\hat{\psi}^{(+)}_{k\prime,k^2}\rangle = 0 \ . \tag{1.63}$$

Since (1.37, 63) give equivalent descriptions of the same two-particle system, they must provide the same values of observables of this system, such as the eigenvalues and scattering phases. For this reason the potentials (1.58) are called phase-equivalent.

The following example demonstrates that not all off-shell continuations are admissible.

Consider, once again, the Lippmann–Schwinger equation for the wave function $|\bar{\psi}\rangle$:

$$|\bar{\psi}^{(+)}_{k\prime\prime,k^2}\rangle = |\bar{k}''\rangle + (k^2 - U^+H_0U + \mathrm{i}0)^{-1}(H - U^+H_0U)|\bar{\psi}^{(+)}_{k\prime\prime,k^2}\rangle \tag{1.64}$$

or

$$|\bar{\psi}^{(+)}_{k\prime\prime,k^2}\rangle = \bar{\Omega}(k^2+\mathrm{i}0)|\bar{k}''\rangle \ . \tag{1.65}$$

We introduce a new wave operator $\tilde{\Omega}$:

$$\tilde{\Omega}(k^2+\mathrm{i}0) = \bar{\Omega}(k^2+\mathrm{i}0)U^+ = U^+\hat{\Omega}(k^2+\mathrm{i}0) \ . \tag{1.66}$$

If now we define the scattering operator $\tilde{T}(k^2)$ by the relation

$$\tilde{T}(k^2) = V\tilde{\Omega}(k^2 + i0) \ , \tag{1.67}$$

it then coincides with the operator $T(k^2)$ on the mass and half-mass shells. This follows from

$$\begin{aligned}\langle k'|\tilde{T}(k^2)|k''\rangle &= \langle k\prime|T(k^2)|k''\rangle \\ &+ (k'^2 - k''^2)\langle k'|V(k^2 - H + i0)^{-1}(U^+ - 1)|k''\rangle\end{aligned} \tag{1.68}$$

which is readily obtained when (1.51,66) are taken into account.

However, it is possible to check that the operator $\tilde{T}$ does not satisfy the Lippmann–Schwinger equation, i.e., a continuation such as (1.67) is not admissible since it yields an incorrect asymptotic of the wave function.

Let us now construct t-matrices for the phase-equivalent potentials V and $\hat{V}$. We use the two-potential formula. Let $T(k^2)$ be the transition operator for the interaction $V_1 + V_2$, i.e.,

$$T(k^2) = (V_1 + V_2)\Omega(k^2 + i0) \ , \tag{1.69}$$

where

$$\Omega(Z) = 1 + (Z - H_0)^{-1}(V_1 + V_2)\Omega(Z) \ . \tag{1.70}$$

We introduce the wave operator for the potential V_1:

$$\Omega_1(Z) = 1 + (Z - H_1)^{-1}V_1 \ . \tag{1.71}$$

Here $H_1 = H_0 + V_1$. Then (1.70) can be rewritten as

$$\Omega(Z) = \Omega_1(Z) + (Z - H_1)^{-1}V_2\Omega(Z) \ . \tag{1.72}$$

Using this equation we obtain for the t-matrix

$$T(k^2) = V_1\Omega_1(k^2 + i0) + [1 + V_1(k^2 - H_1 + i0)^{-1}]V_2\Omega(k^2 + i0) \tag{1.73}$$

or

$$T(k^2) = T_1(k^2) + \Omega_1^+(k^2 - i0)V_2\Omega(k^2 + i0) \ . \tag{1.74}$$

Assume that in (1.74)

$$V_1 = V \ ; \quad V_2 = \hat{V} - V \ . \tag{1.75}$$

Then we have

$$H_1 = H_0 + V = H$$

and

$$\hat{T}(k^2) = T(k^2) + \Omega^+(k^2 - i0)(\hat{H} - H)\hat{\Omega}(k^2 + i0) \ , \tag{1.76}$$

where the following notation is introduced

$$\hat{\Omega}(Z) = 1 + (Z - H)^{-1}\hat{V} ; \tag{1.77}$$

$$\hat{H} = UHU^{+} . \tag{1.78}$$

We now express $\hat{\Omega}$ through Ω. It is easy to verify that from (1.52, 59) and from the relation for the off-shell function

$$|\hat{\psi}^{(+)}_{k',k^2}\rangle = \hat{\Omega}(k^2 + i0)|k'\rangle ,$$

one can obtain

$$\hat{\Omega}(k^2 + i0)|k'\rangle = U|\bar{\psi}^{(+)}_{k',k^2}\rangle . \tag{1.79}$$

From (1.44, 54, 79) we can also obtain an expression for $\hat{\Omega}$ in terms of Ω:

$$\hat{\Omega}(Z) = U\Omega(Z) + U(Z - H)^{-1}(U^{+} - 1)(Z - H_0) . \tag{1.80}$$

By substituting (1.80) into (1.76) we get the desired relation between the transition operators $\hat{T}$ and T generated by the phase-equivalent potentials $\hat{V}$ and V:

$$\begin{aligned} \hat{T}(k^2) = T(k^2) &+ \Omega^{+}(k^2 - i0)(U^{+} - 1)(k^2 - H_0) \\ &+ (k^2 - H_0)(U - 1)\Omega(k^2 + i0) \\ &+ (k^2 - H_0)(U - 1)(k^2 - H + i0)^{-1}(U^{+} - 1)(k^2 - H_0) . \end{aligned} \tag{1.81}$$

Note that the condition that the norm of the vector $(U^{+} - 1)|k\rangle$ be finite is sufficient for satisfying

$$\lim_{k \to k'} (k^2 - k'^2)\langle k'|(U - 1)\Omega(k^2 + i0)|k\rangle = 0 . \tag{1.82}$$

Bearing this in mind, we immediately verify that on the mass shell, as was to be expected, $\hat{T}(k^2) = T(k^2)$. For the difference between the completely off-shell matrix elements of the transition operators $\hat{T}$ and T we ultimately obtain

$$\begin{aligned} \langle k'|\hat{T}(k^2)|k''\rangle - \langle k'|T(k^2)|k''\rangle &= (k^2 - k''^2)\langle k'|\Omega^{+}(k^2 - i0)(U^{+} - 1)|k''\rangle \\ &+ (k^2 - k'^2)\langle k'|(U - 1)\Omega(k^2 + i0)|k''\rangle \\ &+ (k^2 - k''^2)(k^2 - k'^2)\langle k'|(U - 1)(k^2 - H + i0)^{-1}(U^{+} - 1)|k''\rangle . \end{aligned} \tag{1.83}$$

1.3 Separable Expansions of the t-Matrix

Let us now consider some approximate methods for solving the two-body problem (for further separable expansions see [5]). We shall be interested in those methods which lead to the approximate T-matrix $t^N_l(k, k', Z)$ which is represented as a sum of terms separable in k and k'. These approximations are extremely convenient in three-body and four-body problems. These approximate methods are also of independent interest, since, to apply them, none of the assumptions usually recommended in quantum mechanics textbooks need to be made.

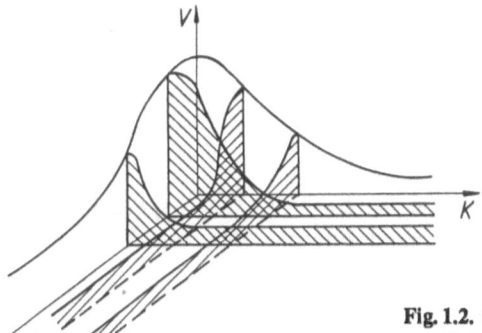

Fig. 1.2. Graphic construction of the approximate potential (1.85)

We shall start with *Bateman's method* [6]. Let there be the S-harmonic of a short-range potential in the momentum space:

$$V(k, k') = \frac{1}{2\pi^2} \int_0^\infty j_0(kr) j_0(k'r) V(r) r^2 dr , \quad j_0(x) = \frac{\sin x}{x} . \tag{1.84}$$

If the potential $V(r)$ is attractive, then for any short-range interaction (i.e., decreasing more rapidly than $1/r^n$) interaction the function $V(k, k')$ has a bell-like shape (Fig. 1.2).

Let us now cut the surface $V(k, k')$ with planes parallel to the coordinate planes. We shall construct the approximate surface $V^N(k, k')$ so that it coincides with the initial surface $V(k, k')$ on the lines of intersection of the intersecting planes with the surface $V(k, k')$.

It is easily verified that an expression of the form

$$V^N(k, k') = \sum_{i,j=1}^N V(k, S_j)(d^{-1})_{ij} V(S_j, k') , \tag{1.85}$$

where $d_{ij} = V(S_i, S_j)$ and S_j are given values of the momenta k and k', satisfies the required property. Indeed, setting in (1.85) $k = S_m (m = 1 \ldots N)$ we have

$$V^N(s_m, k') = \sum_{i,j=1}^N V(s_m s_i)(d^{-1})_{ji} V(s_j, k') ,$$

but because $\sum_j V(S_m, S_j)(d^{-1})_{ij} = \delta_{mj}$,

consequently,

$$V^N(s_m, k') = \sum_{j=1}^{N} \delta_{mj} V(s_j, k') = V(s_m, k') , \qquad (1.86)$$

i.e., the approximate surface $V^N(k, k')$ coincides with the exact one on the line $k = S_m$.

We shall explain (1.85) by using an example. Take the Yukawa potential $V(r) = G \exp(-\mu r)/r$, the S-harmonic of which (1.84) is

$$V(k, k') = \frac{G}{2kk'} \ln \frac{\mu^2 + (k + k')^2}{\mu^2 + (k - k')^2} . \qquad (1.87)$$

As a function of k and k', the potential in (1.87) has the shape depicted in Fig. 1.2. If in (1.85) one chooses $S_1 = 0$ and only one term is included in the sum, then we obtain

$$V^1(k, k') = \frac{G}{(\mu^2 + k^2)(\mu^2 + k'^2)} . \qquad (1.88)$$

Expression (1.88) is the well-known factorized Yamaguchi potential [7], which is widely applied in nuclear physics. Thus, the non-local Yamaguchi potential can be considered as the first approximation to the local Yukawa potential.

Solving the Lippmann–Schwinger equation for the pair t-matrix with the approximate potential (1.85), we easily obtain

$$t^N(k, k', Z) = \sum_{i,j=1}^{N} V(k, s_i)[c^{-1}(Z)]_{ji} V(s_j, k') , \qquad (1.89)$$

where

$$c_{ij}(Z) = d_{ij} + 4\pi m \int_0^\infty \frac{V(k, s_i)V(k, s_j)}{k^2 - mZ} k^2 dk , \qquad (1.90)$$

i.e., for the t-matrix we get an expression which is also separable in the variables k and k'. Obviously, (1.89) yields for the scattering length

$$a \sim \sum_{i,j=1}^{N} V(0, s_i)[C^{-1}(0)]_{ij} V(s_j, 0) .$$

It is sometimes convenient to separate the potential term in (1.89). Simple computations lead to

$$t^N(k, k', Z) = V(k, k') + \sum_{j,i=1}^{N} V(k, s_i)[C^{-1}(Z)]_{ij} l_j(k') , \qquad (1.89')$$

where

$$l_j(k') = \int_0^\infty V(s_i, q) G_0(q, Z) V(q, k') q^2 dq .$$

Until now we have dealt with the approximation of a partial harmonic of the potential $V_l(k, k')$, i.e., with a function of the two variables k and k'. However, in nuclear and atomic physics, situations are often encountered when the system does not possess spherical symmetry. In such cases it is impossible to separate the angular variables in the Lippmann–Schwinger equation by going to partial expansions of the potential and of the T-matrix. Examples of such situations are the following: (a) a system of two particles interacting by means of a non-central (for instance, an axially symmetric) potential; (b) two particles moving with respect to each other with large momentum. In both cases expansion in partial waves would require taking a large number of terms of the expansion into account. Application of the multidimensional generalization of (1.85) makes it possible to do without the partial expansion, however [8].

Consider motion of a particle in an axially symmetric potential. Because of the axial symmetry one can transform from three-dimensional equations to two-dimensional ones by using an expansion of the form

$$\langle \mathbf{k}|V|\mathbf{k}'\rangle = \sum_{m=-\infty}^{\infty} V_m(\tilde{k}, \tilde{k}') \exp(im\varphi) , \tag{1.91}$$

$$\langle \mathbf{k}|t(Z)|\mathbf{k}'\rangle = \sum t_m(\tilde{k}, \tilde{k}', Z) \exp(im\varphi) , \tag{1.92}$$

where $\tilde{k}$ and $\tilde{k}'$ are two-dimensional vectors; φ is the difference between the azimuthal angles of the vectors $\mathbf{k}$ and $\mathbf{k}'$.

For the components t_m we obtain a two-dimensional Lippmann–Schwinger equation:

$$t_m(\tilde{k}, \tilde{k}', Z) = V_m(\tilde{k}, \tilde{k}') + 2\pi \int d\tilde{q} V_m(\tilde{k}, \tilde{q}) G_0(q^2, Z) t_{m'}(\tilde{q}, \tilde{k}', Z) . \tag{1.93}$$

Substituting the exact non-central potential by an approximate, also non-central, potential of the form

$$V_m^{(N)}(\tilde{k}, \tilde{k}') = \sum_{i,j=1}^{N} \eta_i^{(m)}(\tilde{k}) \left[(d^{(m)})^{-1} \right]_{ij} \eta_j^{(m)}(\tilde{k}') , \tag{1.94}$$

we obtain multidimensional analogues of (1.89, 90):

$$t_m^N(\tilde{k}\tilde{k}', Z) = \sum_{i,j=1}^{N} \eta_i^{(m)}(\tilde{k}) \left[(c^{(m)}(Z))^{-1} \right]_{ij} \eta_j^{(m)}(\tilde{k}') , \tag{1.95}$$

where

$$c_{ij}^{(m)}(Z) = d_{ij}^{(m)} - 2\pi \int d\tilde{q}\,\eta_i^{(m)}(\tilde{q})G_0(q^2, Z)\eta_j^{(m)}(\tilde{q}) ,$$

$$\eta_j^{(m)}(\tilde{k}) = V_m(\tilde{k}, \tilde{s}_j); d_{ij}^{(m)} = V_m(\tilde{s}_i, \tilde{s}_j) ;$$

$\tilde{s}_j$ is a set of fixed two-dimensional vectors.

We now give a real example of the application of this generalization of the Bateman method. We are interested in the spectrum of levels in the non-central potential:

$$V(r) = V_0 \exp(-\alpha r^2 \cos^2 \theta - \beta r^2 \sin^2 \theta) . \tag{1.96}$$

For the m-th harmonic of the Fourier transform of this potential, we have

$$\begin{aligned}V_m(\tilde{k}, \tilde{k}') =&V_0/(8\pi^{3/2}\beta\sqrt{\alpha})\exp[-(k\cos\theta - k'\cos\theta')^2/(4\alpha)\\&- (k\sin\theta - k'\sin\theta')^2/(4\beta)] - \exp(-y)I_m(y) ,\end{aligned} \tag{1.97}$$

where $I_m(y)$ is the Bessel function of an imaginary argument

$$y = kk' \sin\theta \sin\theta'/(2\beta) .$$

The parameters of potential (1.96) are chosen in such a way that in the corresponding central potential (i.e. for $\alpha = \beta = \alpha_0$) there exist four levels: $1S, 1P, 2S, 2D$. In addition, for convenience, we impose the condition that the volume in the central and non-central potentials be equal to each other, that is, $\alpha\beta^2 = \alpha_0^3$.

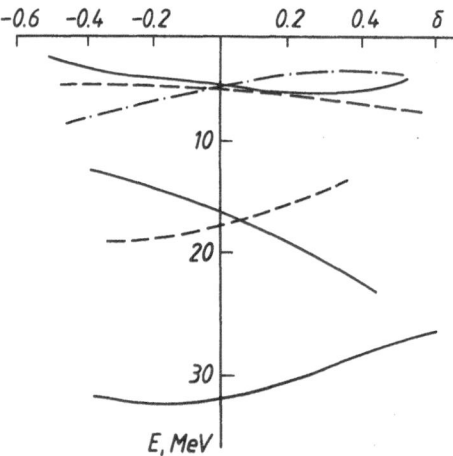

Fig. 1.3. Dependence of the energy levels on the deformation δ: the splitting of levels with respect to the projection m is seen. This is characteristic of non-central potentials

If one introduces the deformation parameter

$$\delta = 1 - \alpha/\beta \tag{1.98}$$

characterizing deviation from central symmetry ($\delta = 0$), then from the condition $\det C^m(E) = 0$, the spectrum $E^{nlm}(\delta)$ depicted in Fig. 1.3 can be obtained. Here

n is the principal quantum number; l is the angular momentum of the level in the central potential. As one can see from this formulation, it permits a departure from perturbation theory in terms of the deformation parameter δ.

We shall now consider another method of approximate solution of the Lippmann–Schwinger equation for the T-matrix based on application of the *Schwinger variational principle*. (A more detailed exposition of the applications of this variational principle and its generalizations can be found in [9].)

Assume a set of linearly independent functions $|f_n\rangle$. We shall demonstrate that it is possible to obtain the approximate expression for the T-matrix

$$t^N = \sum_{m,n=1}^{N} V|f_m\rangle (D^{-1})_{mn} \langle f_n|V \,, \tag{1.99}$$

where

$$D_{mn} = \langle f_n|(V - VG_0(E)V)|f_m\rangle \,; \tag{1.100}$$

V is the pair potential.

The Schwinger functional for the T-matrix is known to be of the form

$$\langle p|t(E)|p'\rangle = \langle p|V|\psi_{p'}^{(+)}\rangle + \langle \psi_p^{(-)}|V|p'\rangle$$
$$- \langle \psi_p^{(-)}|(V - VG_0(E)V)|\psi_{p'}^{(+)}\rangle \,. \tag{1.101}$$

Starting from the Lippmann–Schwinger equations for the wave functions $|\psi^{(+)}\rangle$ and $\langle \psi^{(-)}|$:

$$|\psi_{p'}^{(+)}\rangle = |p'\rangle + G_0(E + i0)V|\psi_{p'}^{(+)}\rangle \,; \tag{1.102}$$

$$\langle \psi_p^{(-)}| = \langle p| + \langle \psi_p^{(-)}|VG_0(E - i0) \,, \tag{1.103}$$

one can readily verify that the functional (1.101) is stable with respect to variation of the functions $|\psi^{(+)}\rangle$ and $\langle \psi^{(-)}|$ about their exact values. As test functions in the functional (1.101) we shall consider linear combinations of some of the functions $|f_n\rangle$:

$$|\psi_{p'}^{(+)}\rangle = \sum_{n=1}^{N} a_n(p')|f_n\rangle \,; \tag{1.104}$$

$$\langle \psi_p^{(-)}| = \sum_{n=1}^{N} \langle f_n|b_n(p) \,. \tag{1.105}$$

By substituting (1.104, 105) into (1.101) and requiring the functional to be stable with respect to variations of the coefficients a and b, we find

$$b_{n'}(p) = \sum_{n} \langle p|V|f_n\rangle (D^{-1})_{nn'} \,; \tag{1.106}$$

$$a_n(p') = \sum_{n'} (D^{-1})_{nn'} \langle f_{n'}|V|p' \rangle \,, \tag{1.107}$$

where

$$D_{mn} = \langle f_m|(V - VG_0(E)V)|f_n \rangle \,. \tag{1.108}$$

Utilizing (1.101, 106–108), we obtain the following expression separable in p and p' for the T-matrix:

$$\langle p|t^N(E)|p' \rangle = \sum_{nn'=1}^{N} \langle p|V|f_n \rangle [D^{-1}(E)]_{nn'} \langle f_n|V|p' \rangle \,. \tag{1.109}$$

It is clear that if the functions $|f_n\rangle$ are taken to be plane waves $|k_n\rangle$, then we obtain the already familiar three-dimensional analogue of the separable Bateman expansion.

In application of these approximate expressions it is convenient to write them for partial waves. We introduce the functions

$$\langle p|pLM \rangle \equiv \sqrt{\frac{\pi}{2} \cdot 2\mu} \frac{\delta(p'-p)}{p^2} Y_{LM}(p) \,, \tag{1.110}$$

where μ is the reduced mass. Then

$$\langle p|p \rangle = \delta(p - p') \,; \tag{1.111}$$

$$\langle p'LM'|pLM \rangle = \frac{\pi}{2} \cdot 2\mu \delta_{LL'} \delta_{MM'} \frac{\delta(p'-p)}{p^2} \tag{1.112}$$

and for a plane wave we obtain

$$|p\rangle = \sqrt{\frac{2}{\pi} \cdot 2\mu} \sum_{LM} |pLM\rangle Y^*_{LM}(\hat{p}) \,. \tag{1.113}$$

In coordinate representation the functions (1.110) have the following form:

$$\langle r|pLM \rangle = \sqrt{2\mu} j_L(pr) i^L Y_{LM}(\hat{r}) \,. \tag{1.114}$$

The equation for the partial amplitudes is

$$\langle p|T_L(Z)|p' \rangle = \langle p|V_L|p' \rangle + \frac{2}{\pi} \int_0^\infty \frac{\langle p|V_L|q \rangle \langle T_L(S)|p' \rangle}{2\mu Z - q^2} q^2 dq \,, \tag{1.115}$$

where on the mass shell

$$\langle k|T_L(Z)|k \rangle = -(1/k)\exp i\delta_L \sin\delta_L \,; \quad Z = k^2/2\mu + i0 \,. \tag{1.116}$$

The total amplitude is

$$\langle p|t(Z)|p'\rangle = \frac{2}{\pi} \cdot \frac{1}{2\mu} \sum_{LM} Y_{LM}(\hat{p})\langle p|T_L(Z)|p'\rangle Y^*_{LM}(\hat{p}') \ . \tag{1.117}$$

For the functions $|f_n\rangle$, we choose the eigenfunctions of the operators L^2 and L_Z (L is the angular momentum), i.e.,

$$\langle p|f_{nLM}\rangle = f_{nL}(p)Y_{LM}(\hat{p}), n = 1,\dots N \ . \tag{1.118}$$

Then

$$[D_L(Z)]_{nn'} = \frac{2}{\pi}\frac{1}{2\mu} \int\limits_0^\infty p'^2 dp' p^2 dp \, f_{n'L}(p')f_{nL}(p)$$

$$\times \left[\langle p'|V_L|p\rangle - \frac{2}{\pi}\int\limits_0^\infty \frac{\langle p'|V_L|p''\rangle\langle p''|V_L|p\rangle}{2\mu Z - p''^2}p''^2 dp'' \right] \tag{1.119}$$

and for the partial t-matrix we obtain

$$\langle p|T_L(Z)|p'\rangle = \sum_{n,n'=1}^N \langle pL|V|f_{nL}\rangle[D^{-1}(Z)]_{nn'}\langle f_{n'L}|V|p'L\rangle \ , \tag{1.120}$$

where the following notation is introduced

$$\langle pL|V|f_{nL}\rangle \equiv \langle pLM|V|f_{nLM}\rangle = \sqrt{\frac{2}{\pi}\frac{1}{2\mu}} \int\limits_0^\infty \langle p|V_L|q\rangle f_{nL}(q)q^2 \, dq \ . \tag{1.121}$$

In the coordinate representation this matrix element is

$$\langle pL|V|f_{nL}\rangle = \sqrt{2\mu} \int\limits_0^\infty j_L(pr)V(r)g_{nL}(r)r^2 dr \ , \tag{1.122}$$

where

$$g_{nL}(r) = \sqrt{\frac{2}{\pi}} \int\limits_0^\infty j_L(pr)f_{nL}(p)p^2 dp \ .$$

As in (1.89), in the approximate expression for the t-matrix (1.120) one can single out the potential as a separate term.

In practical applications of (1.89, 120) problems can arise when choosing the parameters s_j in the Bateman expansion and the form of the functions $f_{nL}(q)$ in expansion (1.120). In the first case, the parameters s_j can be found from the condition that the functional $\Phi(s_1 \dots s_N)$ be minimized with respect to these parameters. Where $\Phi(s_1 \dots s_N)$ has the form

$$\Phi(s_1 \ldots s_N) = \int\limits_0^\infty |V(k, k') - V^{(N)}(k, k', s_1 \ldots s_N)|^2 dk \, dk' \; .$$

For example, for the Yukawa potential in which there exists only one bound state, it suffices to take only three parameters s_j so as to achieve the condition

$$\varepsilon = \min_{s_j} \Phi(s_1 \ldots s_N)/ \int |V|^2 dk \, dk' \approx 10^{-3} \; . \tag{1.123}$$

Thus, it is clear that condition (1.123) is of a purely "geometrical" nature. It reflects the degree of closeness of the approximate surface $V^N(k, k')$ to the exact one $V(k, k')$, and does not require the potential $V(k, k')$ to be small with respect to anything, nor does it require any particular values of the system energy since the functional Φ is independent of energy.

As regards the choice of the functions $f_{nL}(q)$, only the most general assumptions can be made. First of all, it is clear that these functions must allow all the quantities entering into (1.120) to exist. This means that $f_{nL}(q)$ should not be orthogonal to the potential, i.e., the integral (1.121) must not be small. In addition, it is clear that these functions must be linearly independent, otherwise the matrix $D_L^{-1}(Z)$ would not exist. Secondly, the decrease of the T-matrix as a power of p at large p, which is established independently, occurs if the function $f_{nL}(q)$ also decreases as a power function. If some a priori information exists on the scattering wave functions, it may also be taken into account in choosing the functions $f_{nL}(q)$ on the basis of (1.104–108).

We now consider the *approximate representation of the t-matrix in a separable form based on the Bubnov–Galerkin method* [10]. We shall deal with a special choice of the functions $f_{nL}(q)$ based on simple physical arguments. We shall assume that we are interested in the relative motion of two particles at low energies, such that the wave function describing, for example, particle scattering, does not oscillate strongly within the range of action of the potential $V(r)$. It is then clear that in this region the wave function can be approximated by a sum of polynomials of low order. We shall now verify that such an approximation of the wave function is equivalent to representing the t-matrix in the separable form (1.120).

Thus, in accordance with the Bubnov–Galerkin method we shall attempt to find the wave function of the system in the form

$$\psi_k(r, Z) = \sum_{n=1}^N C_n(k, Z)\varphi_n(r) \; , \tag{1.124}$$

where for $\varphi_n(r)$ we shall take polynomials of the order n, normalized with a weighting factor $V(r)$, i.e.,

$$\int\limits_0^\infty r^2 dr \, V(r)\varphi_n(r)\varphi_m(r) = \delta_{mn} \; . \tag{1.125}$$

(For simplicity we present the procedure for a potential of constant sign, i.e., either purely attractive, or purely repulsive. The generalization to potentials of alternating sign is given in [10].) Such a set of polynomials can always be constructed, if the integral $\int_0^\infty r^2 dr\, V(r)$ exists.

We now take advantage of the S-wave harmonic of (1.7) in the coordinate representation:

$$\psi_k(r, Z) = j_0(kr) + \frac{2}{\pi} \int_0^\infty r'^2 dr'\, K(r, r', Z) V(r') \psi_k(r', Z) \,, \tag{1.126}$$

where

$$K(r, r', Z) = \int_0^\infty p^2 dp \frac{j_0(pr) j_0(pr')}{Z - p^2/2\mu} \,, \tag{1.127}$$

as well as the S-harmonic of the expression for the *t*-matrix (1.8):

$$t(p, k, Z) = \int_0^\infty r^2 dr\, j_0(pr) V(r) \psi_k(r, Z) \,. \tag{1.128}$$

Combining (1.124–128) for the *t*-matrix we obtain

$$t^N(p, k, Z) = \sum_{mn=0}^N M_m(p) [A^{-1}(Z)]_{mn} M_n(k) \,, \tag{1.129}$$

where

$$M_n(p) = \int_0^\infty r^2 dr\, \varphi_n(r) V(r) j_0(pr) \,; \tag{1.130}$$

$$A_{mn}(Z) = \delta_{mn} - B_{mn}(Z)$$
$$B_{mn}(Z) = \frac{2}{\pi} \int_0^\infty p^2 dp \frac{M_m(p) M_n(p)}{p^2/2\mu - Z} \,. \tag{1.131}$$

It is easy to verify that (1.129) can be obtained from the Lippmann–Schwinger equation for the *t*-matrix if the following assumption for the potential is made:

$$V(p, k) \approx V^N(p, k) = \sum_{n=0}^N M_n(p) M_n(k) \,. \tag{1.132}$$

The separable potential (1.132) also approximates well the initial local potential $V(r)$ [10].

Let us now deal with the separable representation of the two-particle T-matrix based on the method of moments [11]. Firstly, we note that expansion (1.124) represents a sum of terms separable in the variables k and r. Bearing in mind the definition of the t-matrix in the form (1.128), we immediately arrive at the conclusion that, having the expansion for the wave function in the form given by (1.124), we shall always obtain the expansion of the t-matrix separable in the variables k and k'. In the method of moments, the solution of the Lippmann–Schwinger equation for the wave function $\psi_k(r, Z)$ is also sought in the form of the following expansion separable in k and r.

$$\psi_k^{(N)}(r, Z) = \sum_{n=0}^{N} C_n^N(k, Z) Y_n(r, Z) . \qquad (1.133)$$

The functions Y_n are moments of the kernel $V G_0$ of the Lippmann–Schwinger equation, for example,

$$Y_n(r, Z) = \int_0^\infty V(r) G_0(r, r') Y_{n-1}(r', Z) r'^2 dr' . \qquad (1.134)$$

The unknown coefficients $C_n^N(k, Z)$ are found from the condition that the deviation of the Lippmann–Schwinger equation be orthogonal to the functions $Y_n(r, Z)$, i.e.,

$$\int_0^\infty [(G_0 V \psi^{(N)})(r) - \psi^{(N)}(r)] Y_\lambda(r, Z) r^2 dr = 0 . \qquad (1.135)$$

As a result we obtain for the $C_n^{(N)}(k, Z)$ a set of algebraic equations

$$\sum_{n'=0}^{N-1} [a_{n,n'}(Z) - a_{n,n'+1}(Z)] C_{n'}^{(N)}(k, Z) = \mu_n(k, Z) , \qquad (1.136)$$

where

$$\mu_n(k, Z) = \int_0^\infty j_0(kr) Y_n(r, Z) r^2 dr ; \quad a_{n,n'} = \int_0^\infty Y_n(r, Z) Y_{n'}(r, Z) r^2 dr .$$

An extremely efficient version of the application of the method of moments for solving the two-particle Lippmann–Schwinger equation was proposed in [12]. Instead of the iterative equation (1.134), the moments of the kernel of the Lippmann–Schwinger equation $V G_0(E)$ were constructed by means of a certain a priori set of linearly independent functions $g_i(k)$:

$$g_i(k) = \frac{1}{k^2 + \beta^2} \left(\frac{k^2}{k^2 + \beta^2} \right)^{i-1} , \quad i = 1 \dots N . \qquad (1.137)$$

Such a choice of test functions accounts for the correct behavior of the S-phase at large k.

Furthermore, a quadrature rule (i.e., weighting factors) has been chosen so that for the moments l_i of the kernel $VG_0(E - i0)$,

$$l_j(p_i, k^2) = \frac{2}{\pi} \int_0^\infty \frac{V(p_i p)g_j(p)p^2 dp}{p^2 - k^2 - i0}$$ (1.138)

the following exact relation is fulfilled:

$$l_j(p_i, k^2) = \sum_{l=0}^N W_{il}(k^2)g_j(p_l) .$$ (1.139)

The weights W_{ij} are then determined from (1.139). For example, setting $p_N = k$ we obtain $\mathrm{Im}\, W_{il}$:

$$\mathrm{Im}\, W_{il}(k^2) = \delta_{lN} k V(p_i, k) .$$ (1.140)

Applying the quadrature rule (1.139) and the expansion of the t-matrix in the functions $g_i(p)$

$$t^{(N)}(p, p', k) = \sum_{i=1}^N C_i(p', k)g_i(p) ,$$ (1.141)

we have for the integral term of the Lippmann–Schwinger equation

$$\frac{2}{\pi} \int_0^\infty \frac{V(p_i, p)t^{(N)}(p, p', k)p^2 dp}{p^2 - k^2 - i\varepsilon} = \sum_{l=1}^N W_{il}(k^2)t^{(N)}(p_l, p', k) .$$ (1.142)

Collecting all the terms in the Lippmann–Schwinger equation we obtain a set of algebraic equations for $t^{(N)}(p_l, p', k)$:

$$\sum_{l=1}^N [\delta_{il} + W_{il}(k^2)]t^{(N)}(p_l, p', k^2) = V(p_l, p') .$$ (1.143)

Combining the solution of the set (1.143) together with (1.141) we find the unknown coefficients of the expansion, $C_i(p', k)$ and thus we complete the derivation of the analytic representation for the approximate solution $t^{(N)}(p, p', k)$. A numerical computation of the half-mass-shell amplitude $f^{(N)}(p, k) = t^{(N)}(p, k, k)$ $/t^{(N)}(k, k, k)$ for the Yukawa potential with triplet parameters at $k = 0.1$ within a wide range of p yields an uncertainty of 1% for $N = 5$ and 0.1% for $N = 7$.

We shall turn to the method of moments again in describing the procedures for an approximate solution of the three-particle equations.

Consider a separable expansion of the pair potential and of the t-matrix in a series of eigenfunctions of the kernel of the Lippmann–Schwinger equation, i.e., in eigenfunctions of the *Hilbert–Schmidt* problem. At first, however, we shall discuss the problem, closely related to such expansions, of improving the convergence of the Born series for the T-matrix [13]. This problem is of significant independent interest for computation of quantities such as the binding energy and scattering phases in a two-particle system.

Let the system Hamiltonian be $H = H_0 + V$ where, as usual, H_0 is the free Hamiltonian and V is the interaction potential. Consider the Lippmann–Schwinger equation for the t-matrix in operator form:

$$T(Z) = V + VG_0(Z)T(Z) .$$

We recall that the Born series of this equation simply represents its expansion in V:

$$T(Z) = V + VG_0(Z)V + \dots . \tag{1.144}$$

Note that none of the terms in (1.144) exhibits a singularity at negative energy. This means that in order to find a pole in the t-matrix corresponding to the existence of a bound state, it is necessary to take into account an infinite number of terms of the Born series. In other words, this series diverges at an energy equal to $-B$, the energy of the bound state of the system. Since divergence is related to a pole in the t-matrix, it makes sense to try to single out the pole explicitly and to construct the Born series for the pole-free part.

Actually, we represent $T(Z)$ in the form

$$T(Z) = \frac{|\psi_1\rangle\langle\psi_1|}{1 - \eta_1(Z)} + T_1(Z) , \tag{1.145}$$

where $T_1(Z)$ has no pole (in this example we assume that there exists a single level in the system) and $|\psi_1\rangle$ and $\eta_1(Z)$ are the eigenfunction and eigenvalue of the kernel of the Lippmann–Schwinger equation, i.e.,

$$VG_0(Z)|\psi_1\rangle = \eta_1(Z)|\psi_1\rangle . \tag{1.146}$$

The eigenvalues for certain $\nu, \eta_\nu(Z)$, possess the following property:

$$\eta_\nu(-B_\nu) = 1 . \tag{1.147}$$

Thus, when (1.147) holds, (1.146) describes the conventional bound-states problem, while the t-matrix (1.145) acquires a pole in the energy at the point $E = -B$.

We now substitute (1.145) into the Lippmann–Schwinger equation; as a result we obtain for T_1 the equation

$$T_1 = V_1 + VG_0T_1 , \tag{1.148}$$

where

$$V_1 = V - |\psi_1\rangle\langle\psi_1| = V - V_s . \tag{1.149}$$

Using (1.149) we exclude V from (1.148) and for T_1, upon inversion of its separable part, we ultimately obtain

$$T_1(Z) = v_1(Z) + v_1(Z)G_0(Z)T_1(Z) , \qquad (1.150)$$

where

$$v_1(Z) = [1 - V_s G_0(Z)]^{-1} V_1 .$$

Thus, proceeding from (1.150) we obtain an iterative procedure in the "reduced" interaction $v_1 \sim V_1$. If several bound states exist, (1.145) becomes

$$T(Z) = \sum_{n=1}^{N} \frac{|\psi_n\rangle\langle\psi_n|}{1 - \eta_n(Z)} + \tilde{T}_1 . \qquad (1.151)$$

Even when a single level is present in the system, (1.151) with $\tilde{T}_1 = 0$ approximates quite well the exact t-matrix for a small number of terms N. Equation (1.151) with $\tilde{T}_1 = 0$ is not unitary for $E > 0$. (It should be noted that the approximation is very poor for Coulomb potentials which we are not considering.)

The fact that it is possible to apply perturbation theory in case of the "reduced" interaction V_1 may also be utilized in solving the Hilbert–Schmidt problem.

In this method it is necessary to perform an independent computation of the eigenfunctions $|\psi_n\rangle$ and the eigenvalues $\eta_n(Z)$ of the Hilbert–Schmidt equation. Regretfully, such solutions are known only for several simple potentials (for instance, the square well, the Hulten potential in the S-state, the Coulomb potential). Nevertheless, the iterative procedure applied in [13] for the Yukawa potential for constructing $|\psi_n\rangle$ has yielded not bad results.

A successful version (1.151) is the expansion in the eigenfunctions $|\psi_n\rangle$ found at a fixed negative energy, for example, the binding energy of a two-particle system. (In the literature this expansion is usually called the unitary-pole expansion [14], since the t-matrix obtained, $t^{(N)}$, is unitary for any number of terms of the expansion, N.)

2. The Faddeev Equations
   in the Three-Body Problem.
   Pion-Nucleus Scattering

This chapter presents a derivation of the three-particle Faddeev equations and of their various modifications as applied in scientific investigations. Other approaches to the three-body problem, such as the method of evolution in the coupling constant and dispersion relations, are also discussed. A number of ways for obtaining approximate solutions of the three-particle equations are considered, illustrating their efficiency by specific examples.

2.1 General Remarks

In this lecture we shall proceed to investigate the *motion of three-particle systems*. We shall use the Faddeev equations or their modifications as the principal dynamic equations. We shall then limit our consideration to short-range pair potentials of particle interaction. (The properties of few-body systems with Coulomb interaction are discussed in detail in [15].)

First of all we note that the equations of motion of a three-particle system are not integrable in the general case, i.e., neither the spectrum nor any other observables of such a system can be represented in a quadrature form as can, for example, the solutions of ordinary differential equations. The reason for this is related to the fact that the number of integrals of motion available to such systems is smaller than the number of dynamic variables necessary to describe such a system.

Consider three particles in the center-of-mass system (Fig. 2.1). Clearly, to describe the positions of these particles the Jacobi variables p and q suffice:

$$p = \frac{m_1 p_2 - m_2 p_1}{m_1 + m_2} \; ; \quad q = \frac{m_3(p_1 + p_2) - (m_1 + m_2)p_3}{m_1 + m_2 + m_3}$$

i.e., six variables. If one is interested in the state of the given total angular momentum of the system L, i.e., if one makes use of the conservation of L, it is possible to single out the angular variables (Euler angles) characterizing the orientation of the triangle $\{p, q\}$ in space. Thus, the existence of the integral of motion L permits one to integrate the equations over three more variables. Generally speaking, there are no more integrals of motion, so one must deal with

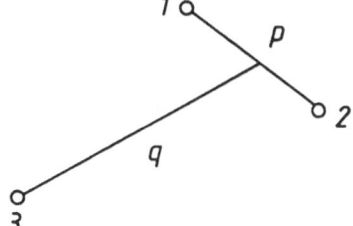

Fig. 2.1. The Jacobi variables for a three-particle system

equations of motion involving three variables. There exist, however, several special cases when it is actually possible to integrate exactly the equations describing three-particle systems. As a rule, such a possibility occurs either when additional integrals of motion, due to the specific features of pair interaction exist, or when constraints are imposed on the system. An example of the first situation is represented by a three-particle system with a pair interaction of the oscillator type. The Hamiltonian of such a system

$$H = \sum_{i=1}^{3} \frac{p_i^2}{2\mu} + \sum_{i<j=1}^{3} V_{ji}$$

is reduced by a simple transformation to

$$H = \sum_{i=1}^{3} \mathcal{H}_i \;,$$

i.e., to the sum of one-particle Hamiltonians, and the wave function may, correspondingly, be written as

$$\Psi = \prod_i \psi_i \;,$$

where $\mathcal{H}_i \psi_i = \varepsilon_i \psi_i$.

An example of an exactly integrable three-particle system with constraints is the well-known Lagrange problem of classical mechanics [16]. In this case the plane motion of three gravitating particles involving a special choice of initial velocities can also be represented in the form of one-particle motion. Another such example is the motion of a single particle in a field with two centers, the distance between which is fixed, while the interaction of the particle with each one of the centers is separable. In this case the solution of the problem for the wave function or scattering amplitude is obtained by quadrature and can be generalized to the case of N centers [17].

Now, let us list the exact results, i.e., the theorems of the theory of three-particle systems. First comes the famous Bruns–Poincaré theorem [18]. This states that in a classical three-particle system there can exist no new algebraic integrals of motion other than the integrals of total energy, total momentum and total angular momentum of the system. (It should be noted that at the time when this theorem was proved only pair interactions of the form κ/r were known.) A subsequent result was given by *Thomas* [19] who observed that a system of three particles with an attractive potential $V(r) = V_0\delta(r)$ collapses, i.e., the spectrum of such a system has no lower bound. A detailed consideration of the properties of such systems is given in [20]. This was followed by L.D. Faddeev's analysis of the non-Fredholm nature of the three-particle Lippmann–Schwinger equations and the formulation of new Fredholm three-body equations – the Faddeev equations – and, finally, the analysis by *V. Efimov* [21] of a special case of the non-Fredholm behavior of three-particle equations, which lead to the discovery of the Efimov effect mentioned in the preface.

Owing to their universal character, the Faddeev equations have found wide application in various branches of physics. In addition to the pure three-particle systems, they have been generalized to describe complexes consisting of three clusters. The development of this generalization has occurred in atomic and nuclear physics. For recent reviews see [22, 23].

One sometimes encounters the opinion that the Faddeev equations can be used only in scattering problems. A simple example can demonstrate that in bound-state problems these equations are even more handy for numerical calculations than the Schrödinger equation. This is because the Faddeev equations are written for the components of the total wave function and, thus, take into account the possible asymptotics of a three-body system.

The difficulties encountered in solving the Faddeev equations should also be pointed out:

As first shown by *Faddeev* [1], when the total energy of a system is positive the kernels of the equations he proposed have "moving" (i.e., dependent on an external variable) logarithmic singularities. They vanish under iterations; however, if one decides to solve the problem without iteration then one must apply special procedures. Below we shall touch upon one such procedure which uses piecewise interpolation of solutions.

Other difficulties are related to the situations when the kernels of the equations cease to be of Fredholm type. Among these are:

1) three-particle systems with a Coulomb pair interaction in the case of positive total energy (renormalization of the Faddeev equations for such systems was accomplished by *Merkuriev* [24]);

2) three-particle systems with a pair potential involving infinite repulsion at short distances (these equations have also been renormalized [25]);

3) three-particle systems in which the Efimov effect is permitted. A qualitative study of the Faddeev equation for this case has been performed by *Faddeev* [26], and independently, by *Amado* and *Noble* [27].

As already pointed out in the preface, the Thomas effect [28] belongs to the same non-Fredholm type.

2.2 Formal Derivations for Transition Amplitudes and Wave Functions

We shall proceed with the formal derivation of the Faddeev equations. Let the total Hamiltonian of a three-particle system be of the form $H = H_0 + V$, where H_0 is the operator of kinetic energy, $V = V_{12} + V_{23} + V_{13}$, and V_{ij} represents the interaction potential between particles i and j.

For the three-particle transition operator T we have the Lippmann–Schwinger equation

$$T(Z) = V + V G_0(Z) T(Z) \tag{2.1}$$

where $G_0(Z) = (Z - H_0)^{-1}$ is the free Green function; $Z = E + \mathrm{i}\varepsilon$.

The kernel of (2.1) is not square integrable, since we are dealing with pair potentials

$$\langle p_1 p_2 p_3 | V | p_1' p_2' p_3' \rangle = \delta(\sum_i p_i - \sum p_i') [\delta(p_3 - p_3') \langle q_{12} | V_{12} | q_{12}' \rangle$$
$$+ \delta(p_1 - p_1') \langle q_{23} | V_{23} | q_{23}' \rangle + \delta(p_2 - p_2') \langle q_{13} | V_{13} | q_{13}' \rangle] . \tag{2.2}$$

Here p_i are the particle momenta, and q_{ij} represents the relative momentum in the pair.

3 ———————————

1 ———————————

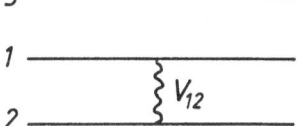 V_{12}

2 ———————————

Fig. 2.2. Disconnected three-particle diagram: particle 3 does not interact in passing, particles 1 and 2 interact by means of the potential V_{12}

We let the diagram in Fig. 2.2 correspond to the first term in the right-hand side of (2.2). Clearly, when (2.1) is iterated, there will always be terms corresponding to multiplication of this diagram by itself and leading to a singularity in the right-hand side of (2.1). To eliminate this defect of (2.1), we introduce, instead of the total operator $T(Z)$, new operators $T^{(j)}$:

$$T(Z) = T^{(1)}(Z) + T^{(2)}(Z) + T^{(3)}(Z) , \tag{2.3}$$

where

$$T^{(k)}(Z) = V_{ij} + V_{ij} G_0(Z) T(Z) , \quad ijk = 123, 231, 312 . \tag{2.4}$$

We now take advantage of the Lippmann–Schwinger equation for pair transition operators

$$T_{ij}(Z) = V_{ij} + V_{ij} G_0(Z) T_{ij}(Z) . \tag{2.5}$$

For this purpose we single out the term $T^{(k)}$ in the right-hand side of (2.4) and transpose it to the left-hand side. Equation (2.4) then assumes the form

$$\left[1 - V_{ij}G_0(Z)\right] T^{(k)}(Z) = V_{ij} + V_{ij}G_0(Z) \left[T^{(i)}(Z) + T^{(j)}(Z)\right] . \tag{2.6}$$

Multiplying (2.6) from the left by the operator $[1 - V_{ij}G_0(Z)]^{-1}$ and utilizing the equation for the pair T-operator (2.5) we obtain the desired equations:

$$T^{(k)}(Z) = T_{ij}(Z) + T_{ij}(Z)G_0(Z) \left[T^{(i)}(Z) + T^{(j)}(Z)\right] ,$$
$$ijk = 123, 231, 312 . \tag{2.7}$$

It is now clear that, owing to the absence of the diagonal term $T^{(k)}$ in the right-hand side of (2.7), the iterated series of this equation will not contain any disconnected terms (i.e., diagrams such as in Fig. 2.1) and, therefore, (this is a necessary but not a sufficient condition, see the discussion of the Efimov effect) the set of equations (2.7) has a unique solution. The equations for the components of the three-particle wave function $\psi^{(k)}$, where $\psi = \psi^{(1)} + \psi^{(2)} + \psi^{(3)}$, may be obtained, for instance, from the equations for the components of the Green functions $G^{(k)}(Z)$:

$$G^{(k)}(Z) = G_{ij}(Z) - G_0(Z) + G_0(Z)T_{ij}(Z) \left[G^{(i)}(Z) + G^i(Z)\right] \tag{2.8}$$

and the definition

$$\psi^{(k)} = \lim_{\varepsilon \to 0} \varepsilon G^{(k)}(E + i\varepsilon)\Phi , \tag{2.9}$$

where Φ is the asymptotic function in one of the channels. For example, for the scattering of particle 1 on a bound state of particles 2 and 3 one may obtain the equations

$$\phi^{(1)} = \Phi_1 + G_0(Z)T_{23}(Z) \left[\psi^{(2)} + \psi^{(3)}\right] ;$$
$$\psi^{(2)} = G_0(Z)T_{31}(Z) \left[\psi^{(3)} + \psi^{(1)}\right] ; \tag{2.10}$$
$$\psi^{(3)} = G_0(Z)T_{12}(Z) \left[\psi^{(1)} + \psi^{(2)}\right] .$$

In the case of the bound-state problem ($Z < 0$), obviously the following set of homogeneous equations holds:

$$\psi^{(1)} = G_0(Z)T_{23}(Z) \left[\psi^{(2)} + \psi^{(3)}\right]$$
$$\psi^{(2)} = G_0(Z)T_{31}(Z) \left[\psi^{(3)} + \psi^{(1)}\right] \tag{2.11}$$
$$\psi^{(3)} = G_0(Z)T_{12}(Z) \left[\psi^{(1)} + \psi^{(2)}\right] .$$

Thus, the rearrangement of the three-particle equations was needed in order to obtain, instead of the non-Fredholm equation (2.1), the set of Fredholm equations (2.7, 10) with a unique solution at a given energy. In this connection, it must be stressed that the Lippmann–Schwinger equation (2.5) for two bodies written for an arbitrary reference frame is also non-Fredholm, since

$$\langle \boldsymbol{p}_1\boldsymbol{p}_2|V_{12}\boldsymbol{p}_1\boldsymbol{p}_2'\rangle = \langle \boldsymbol{p}\boldsymbol{q}_{12}|V_{12}|\boldsymbol{p}'\boldsymbol{q}_{12}'\rangle = \delta(\boldsymbol{p} - \boldsymbol{p}')\langle \boldsymbol{q}_{12}|V_{12}|\boldsymbol{q}_{12}'\rangle , \tag{2.12}$$

where p is the momentum of the center of mass of the pair 1,2. In this case, however, the singularity is eliminated in a trivial way by transition to the center-of-mass system of the pair, as a result of which we obtain an equation with a good kernel with respect to the relative variable q_{12}, as it is well known.

We shall now derive the Faddeev equations in the form in which their author proposed them [1].

We start with the Lippmann–Schwinger equation for the three-particle Green function

$$G(Z) = G_0(Z) + G_0(Z)VG(Z) .\tag{2.13}$$

Instead of the function $G(Z)$, we introduce the matrix $M_{\alpha\beta}(Z)$:

$$M_{\alpha\beta}(Z) = \delta_{\alpha\beta}V_\alpha + V_\alpha G(Z)V_\beta .\tag{2.14}$$

Here α is a double index running through the values 2,3, 3,1, and 1,2. One can readily verify that the three-particle transition operator $T(Z)$ is expressed through the operators $M_{\alpha\beta}(Z)$ as follows:

$$T(Z) = \sum_{\alpha\beta} M_{\alpha\beta}(Z) .$$

By using (2.13) one can show that these operators satisfy the set of equations

$$M_{\alpha\beta}(Z) = \delta_{\alpha\beta}V_\alpha + V_\alpha G_0(Z) \sum_\gamma M_{\gamma\beta}(Z) ,\tag{2.15}$$

which, however, is in no way better than (2.13), since the kernels of this set of equations contain the same singularities as does the kernel of (2.13). To eliminate the singularities, we separate the sum in the right-hand side of (2.15) into two terms:

$$V_\alpha G_0(Z)M_{\alpha\beta}(Z) + V_\alpha G_0(Z) \sum_{\gamma \neq \alpha} M_{\gamma\beta}(Z)$$

and transpose the first term to the left-hand side of (2.15):

$$[1 - V_\alpha G_0(Z)] M_{\alpha\beta}(Z) = \delta_{\alpha\beta}V_\alpha + V_\alpha G_0(Z) \sum_{\gamma \neq \alpha} M_{\gamma\beta}(Z) .\tag{2.16}$$

Once again, using the equation for the pair operator $T_\alpha(Z)$

$$\begin{aligned} T_\alpha(Z) &= V_\alpha + V_\alpha G_0(Z)T_\alpha(Z) = V_\alpha + T_\alpha(Z)G_0(Z)V_\alpha \\ &= [1 + T_\alpha(Z)G_0(Z)] V_\alpha \end{aligned}\tag{2.17}$$

and multiplying (2.16) by $1 + T_\alpha(Z)G_0(Z)$ we obtain, taking account of (2.17),

$$M_{\alpha\beta}(Z) = \delta_{\alpha\beta}T_\alpha(Z) + T_\alpha(Z)G_0(Z) \sum_{\gamma \neq \alpha} M_{\gamma\beta}(Z) .\tag{2.18}$$

The set of equations (2.18) is precisely the set of Faddeev equations for the operators $M_{\alpha\beta}(Z)$. If one goes from the operator form of (2.18) to the momentum space representation, it becomes clear that the inhomogeneous term contains a δ-function, owing to its pairlike nature. In this connection, it is sometimes more convenient to replace the operators $M_{\alpha\beta}$ by the iterated operators $W_{\alpha\beta}(Z) = M_{\alpha\beta}(Z) - \delta_{\alpha\beta}T_\alpha(Z)$ for which the following set of equations holds:

$$W_{\alpha\beta}(Z) = W^{(0)}_{\alpha\beta}(Z) + T_\alpha(Z)G_0(Z) \sum_{\gamma \neq \alpha} W_{\gamma\beta}(Z) , \qquad (2.19)$$

where

$$W^{(0)}_{\alpha\alpha}(Z) = 0 ; \quad W^{(0)}_{\alpha\beta}(Z) = T_\alpha(Z)G_0(Z)T_\beta(Z) .$$

To solve the Faddeev equations it is necessary to transform the operator form to some other representation – the momentum space or configurational representation. In this lecture we shall always utilize the momentum representation and the integral form of equations natural for this representation, although the configuration representation, together with the related differential form of the Faddeev equations, has lately received wide acceptance [29]. The differential form is especially convenient for dealing with systems with Coulomb interaction, since it does not involve the pair T-matrix which is quite complicated for such a potential.

As stated above, we shall be interested only in systems with short-range potentials. Such potentials, as a rule [6], can be quite well approximated by a small number of separable potentials, leading to a significant simplification of the integral equations. In addition the pair T-matrix can be found in an extremely simple fashion.

Thus, we write the Faddeev equations (2.10) for the wave function in the momentum representation neglecting, for simplicity, the discrete variables (spin, isospin). The kernels of these equations are determined by the free Green function and the pair T-matrix, for which we have the following matrix elements in the space of three-particle states (plane waves):

$$\langle k'_{jk}p'_i|G_0(Z)k_{jk}p\rangle = \frac{\delta(k'_{jk} - k_{jk})\delta(p - p'_i)}{Z - k^2_{jk}/2\mu_{jk} - p^2_i/2\mu_i} ; \qquad (2.20)$$

$$\langle k'_{jk}p'_i|T_{jk}(Z)|k_{jk}p_i\rangle = \langle k'_{ik}|t_{jk}\left(Z - \frac{p^2_i}{2\mu_i}\right)|k_{jk}\rangle\delta(p_i - p'_i) . \qquad (2.21)$$

Here, once again, k_{jk} and p_i are the Jacobi variables, and $\mu_{jk} = m_im_k/(m_j + m_k)$; $\mu_i = m_i(m_j + m_k)/(m_i + m_j + m_k)$ (Fig. 2.3). The energy argument of the pair T-matrix is as indicated in (2.21), since the energy of the pair subsystem E_{jk}, on which the pair T-matrix $t_{jk}(E_{jk})$ depends, is equal to the difference between the total energy of the system Z and the energy of the third particle, $E_{jk} = Z - p^2_i/2\mu_i$.

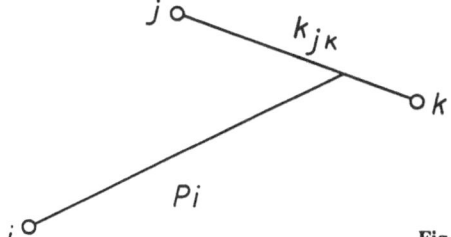

Fig. 2.3. The Jacobi variables for particles i, j, and k

Thus, the Faddeev equations for the wave function in momentum representation are

$$\psi = \psi^{(1)}(k_{23}, p_1) + \psi^{(2)}(k_{31}, p_2) + \psi^{(3)}(k_{12}, p_3) \; ;$$

$$\psi^{(1)}(k_{23}, p_1) = \Phi_1(k_{23}, p_1) + \left(Z - \frac{k_{23}^2}{2\mu_{23}} - \frac{p_1^2}{2\mu_1} \right)^{-1} \int \frac{dk_{23}'}{(2\pi)^3}$$

$$\times \langle k_{23} | t_{23} \left(Z - \frac{p_1^2}{2\mu_1} \right) | k_{23}' \rangle \left[\psi^{(2)} \left(k_{31}', p_2' \right) \right.$$

$$\left. + \psi^{(3)} \left(k_{12}', p_3' \right) \right] \; ;$$

$$\psi^{(2)}(k_{31}, p_2) = \left(Z - \frac{k_{31}^2}{2\mu_{31}} - \frac{p_2^2}{2\mu_2} \right)^{-1} \int \frac{dk_{31}'}{(2\pi)^3}$$

$$\times \langle k_{31} | t_{31} \left(Z - \frac{p_2^2}{2\mu_2} \right) | k_{31}' \rangle \left[\psi^{(3)} \left(k_{12}', p_3' \right) \right. \qquad (2.22)$$

$$\left. + \psi^{(1)} \left(k_{23}', p_1' \right) \right] \; ;$$

$$\psi^{(3)}(k_{12}, p_3) = \left(Z - \frac{k_{12}^2}{2\mu_{12}} - \frac{p_3^2}{2\mu_3} \right)^{-1} \int \frac{dk_{12}'}{(2\pi)^3} \langle k_{12} | t_{12} \left(Z - \frac{p_3}{2\mu_3} \right) | k_{12} \rangle$$

$$\times \left[\psi^{(1)}(k_{23}', p_1') + \psi^{(2)}(k_{31}', p_2') \right] \; ;$$

$$Z = \frac{p_1^{0^2}}{2\mu_1} - \frac{\kappa_{23}^2}{2\mu_{23}} + i\varepsilon \; ,$$

where the following relationships hold: $k_{31}' = -\frac{3}{4}p_1 - \frac{1}{2}k_{23}'$ (for particles of equal mass); $p_1' = -\frac{1}{2}p_1 + k_{23}'$; $k_{12}' = \frac{3}{4}p_1 - \frac{1}{2}k_{23}'$; $p_3' = -\frac{1}{2}p_1 - k_{23}'$.

If all three particles are identical and possess no spin and no isospin, then all three components $\psi^{(j)}$ exhibit the same functional dependence on their variables, such as, for example, $\chi(k, p)$, and the total function of the system is determined by

$$\psi = \chi(k_{23}, p_1) + \chi(k_{31}, p_2) + \chi(k_{12}, p_3) \; ,$$

while for the function χ, from (2.22) we obtain the equation

$$\chi(\boldsymbol{k},\boldsymbol{p}) = \varphi(\boldsymbol{k},\boldsymbol{p}) + \left(Z - \frac{k^2}{m} - \frac{3}{4}\frac{p^2}{m}\right)^{-1} \int \frac{d\boldsymbol{p}'}{(2\pi)^3}$$

$$\times \left\{ \langle \boldsymbol{k}|t\left(Z - \frac{3}{4}\frac{p^2}{m}\right)|\frac{\boldsymbol{p}}{2} + \boldsymbol{p}'\rangle \right. \tag{2.23}$$

$$\left. + \langle \boldsymbol{k}|t\left(Z - \frac{3}{4}\frac{p^2}{m}\right)|-\frac{\boldsymbol{p}}{2} - \boldsymbol{p}'\rangle \right\} \chi\left(\boldsymbol{p} + \frac{\boldsymbol{p}'}{2},\boldsymbol{p}'\right),$$

where $\varphi(\boldsymbol{k},\boldsymbol{p}) = (2\pi)^3 \delta(\boldsymbol{p} - \boldsymbol{p}')\varphi_{\mathrm{d}}(\boldsymbol{k})$.

If the particles have spin and the interaction depends on the spin this, generally speaking, leads to an increase in the number of components of the system wave function.

2.3 Scattering of a Particle on a Bound Pair

We shall now obtain an approximate solution of the multidimensional equation (2.23). Our aim will be to reduce to a one-dimensional equation which is readily solved using computer methods. For simplicity we consider three identical spinless bosons.

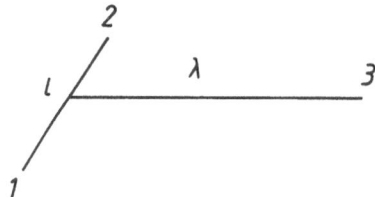

Fig. 2.4. First term of the right-hand side of (2.2): l is the angular momentum of particle 3 relative to the center of mass of particles 1 and 2

First, we shall single out the angular variables. We introduce the angular momenta l and λ, the meaning of which becomes clear from Fig. 2.4. The sum $L = l + \lambda$ obviously represents the total angular momentum of the system. We now construct the eigenfunctions of the operators L^2 and M (M is the projection of L onto the Z-axis):

$$\mathcal{Y}_{l\lambda}^{LM}(\boldsymbol{n}_k,\boldsymbol{n}_p) = \sum_{m\mu}(lm\lambda\mu|LM)Y_{lm}(\boldsymbol{n}_k)Y_{\lambda\mu}(\boldsymbol{n}_p). \tag{2.24}$$

Here $\boldsymbol{n}_k$ and $\boldsymbol{n}_p$ are unit vectors directed along the vectors $\boldsymbol{k}$ and $\boldsymbol{p}$, while $(lm\lambda\mu|LM)$ is the Clebsch–Gordon coefficient.

We recall that the vectors $\boldsymbol{k}$ and $\boldsymbol{p}$ occurring in the wave function $\chi(\boldsymbol{k},\boldsymbol{p})$ are dynamic variables. Besides these vectors, the function $\chi(\boldsymbol{k},\boldsymbol{p})$ contains a dependence on the vector $\boldsymbol{p}_0$, the momentum of the incident particle. In this connection, the partial expansion for the function $\chi(\boldsymbol{k},\boldsymbol{p},\boldsymbol{p}_0)$ has the form

$$\chi(\boldsymbol{k},\boldsymbol{p},\boldsymbol{p_0}) = \sum_{l\lambda LM} \chi_{l\lambda L}(k,p,p_0)\mathcal{Y}_{l\lambda}^{LM}(\boldsymbol{n_k},\boldsymbol{n_p})Y_{LM}^*(\boldsymbol{n_{p_0}}) \ . \tag{2.25}$$

Substituting (2.25) and $\langle\boldsymbol{k'}|t(Z)|\boldsymbol{k}\rangle = \sum_l(2l+1)t_l(k',k,Z)P_l(\cos\theta)$ into (2.23) we obtain for $\chi_{l\lambda L}$ the following set of two-dimensional equations (under the assumption that the angular momentum of the bound pair equals zero):

$$\chi_{l\lambda L}(k,p,p_0) = (2\pi)^3\varphi_{10}(k)\frac{\delta(p-p_0)}{p^2}\delta_{l0}\delta_{\lambda L}$$

$$+ 2\Delta_l\left(Z_p - \frac{k^2}{m}\right)^{-1}\frac{1}{2\pi^2}\int dp'$$

$$\times\int dn_p\mathcal{Y}_{l\lambda}^{*L^0}(\boldsymbol{n_q},\boldsymbol{n_p})t_l(q,k,Z_p)$$

$$\times\sum_{l'\lambda'}\chi_{l'\lambda'L}(q',p',p_0)\mathcal{Y}_{l'\lambda'}^{L^0}(\boldsymbol{n_{q'}},\boldsymbol{n_{p'}}) \ , \tag{2.26}$$

where $Z_p \equiv Z - \frac{3}{4}\frac{p^2}{m}$; $\Delta_l = [1 + (-1)^l]\frac{1}{2}$; $q = \frac{p}{2} + p'$, $q' = p + \frac{p'}{2}$.

We shall restrict ourselves to the case $l = \lambda = L = 0$. This approximation generally works well when the energy of the incident particle is not high. Equation (2.26) then becomes the two-dimensional integral equation

$$\chi(k,p,p_0) = (2\pi)^3\varphi_{10}(k)\frac{\delta(p-p_0)}{p^2}$$

$$+ \left(Z_p - \frac{k^2}{m}\right)^{-1}\frac{1}{2\pi^2}\int_0^\infty dp'\int_{|p-\frac{p'}{2}|}^{p+p'/2} dk'$$

$$\times\frac{k'p'}{p}t_0(q,k,Z_p)\chi(k',p',p_0) \ , \tag{2.27}$$

$$q = \sqrt{k'^2 + \frac{3}{4}p'^2 - \frac{3}{4}p^2} \ .$$

We shall now take advantage of the separable approximation of the pair t-matrix by the Bateman method with $N = 1$. From (1.89) we have for the t-matrix

$$t(q,k,Z_p) = V(q,s_1)V(s_1,k)/C(Z_p) \ . \tag{2.28}$$

Substituting (2.28) into (2.27) we obtain

$$\chi(k,p) = (2\pi)^3 \varphi_{10}(k)\frac{\delta(p-p_0)}{p^2} + \left(Z_p - \frac{k^2}{m}\right)^{-1}$$

$$\times \frac{1}{2\pi^2}\int_0^\infty dp' \int_{|p-\frac{p'}{2}|}^{p+\frac{1}{2}p'} dk' \frac{k'p'}{p} \frac{V(q,s_1)}{C(Z_p)}$$

$$\times V(k,s_1)\chi(k',p') = (2\pi)^3 \varphi_{10}(k)\frac{\delta(p-p_0)}{p^2}$$

$$+ \frac{V(k,s_1)}{\left(Z_p - \frac{k^2}{m}\right)C(Z_p)}\frac{1}{2\pi^2}\underbrace{\int_0^\infty dp' \int_{|p-\frac{p'}{2}|}^{p+p'/2} dk' \frac{k'p'}{p} V(q,S_1)\chi(k',p')}_{B(p)} .$$

Thus, for the function $\chi(k,p)$ we obtain

$$\chi(k,p) = (2\pi)^3 \varphi_{10}(k)\frac{\delta(p-p_0)}{p^2} + \frac{V(k,s_1)B(p)}{(Z_p - k^2/m)C(Z_P)} , \tag{2.29}$$

where $B(p)$ is an unknown function of a single variable. Using (2.29) in the definition of the function $B(p)$ we get the desired one-dimensional equation

$$B(p) = B_0(p) + \frac{1}{2\pi^2}\int_0^\infty dp' \frac{K(p,p')}{C(Z_{p'})} B(p') , \tag{2.30}$$

where

$$B_0(p) = \frac{4\pi}{pp_0}\int_{|p-\frac{p_0}{2}|}^{p+p_0/2} k' dk' V(q^0, s_1)\varphi_{10}(k') ; \quad q^0 = \sqrt{k'^2 + \frac{3}{4}(p_0^2 - p^2)} ,$$

and

$$K(p,p') = \int_{|p-p'/2|}^{p+p'/2} dk' \frac{k'p'}{p} \frac{V(q,s_1)V(k',s_1)}{Z_{p'} - k'^2/m} . \tag{2.31}$$

2.4 Interpolation Method in the Solution of the Faddeev Equations

We shall now deal with the solution of the three-particle equations [30] in the case when the *total energy of the system is positive*, i.e. we shall consider the situation when, in addition to elastic scattering, breakup of the target can occur. As already noted, moving logarithmic singularities arise in the kernels of the Faddeev equations at such energies. In order to draw special attention to these singularities and to the ways of overcoming them and also, to not encumber the text with too many details, we shall consider the simplified version of the Faddeev equations, the so-called Skornyakov–Ter–Martirosyan equations (S.-T.) [31]. These equations describe scattering in a system consisting of three particles with a delta-like pair potential.

Thus, the S.-T. equation describing quartet nd S-scattering has the form

$$\frac{\sqrt{(3/4)k^2 - Z} - \alpha}{k^2 - k_0^2} a(k, k_0)$$

$$= -\frac{G_0(k, k_0)}{2kk_0} - \frac{2}{\pi} \int_0^\infty \frac{G_0(k, k')a(k', k_0)k'^2 dk'}{2kk'(k'^2 - k_0^2 - i0)} , \tag{2.32}$$

where $a(k, k_0)$ is the S-component of the nd quartet scattering amplitude;

$$G_0(k, k') = \int_{-1}^{1} \frac{dx}{Q + x - i0} , \quad Q = \frac{k^2 + k'^2 - E}{kk'} ; \quad E = -a^2 + \frac{3}{4}k_0^2 ;$$

E is the total energy of the system ; $Z = E + i0$; $\alpha^2 = m|\varepsilon_d|$; $k_0^2 = \frac{3}{4}mE_n$.

Clearly, $\frac{G_0(k,k')}{2kk'}$ is just the angular integral of the free Green function. For convenience we introduce in (2.32) the dimensionless variables:

$$x = k/\sqrt{(4/3)E} ; \quad y = k'/\sqrt{(4/3)E} ; \quad x_0 = k_0/\sqrt{(4/3)E} ;$$
$$\gamma = \alpha/\sqrt{(4/3)E} ,$$

and, instead of the amplitude $a(k, k_0)$, we use

$$B(x) = \frac{a(x, x_0)\sqrt{E}}{\left[(2\gamma/\sqrt{3}) + \sqrt{x^2 - 1}\right][1 - ix_0 a(x_0, x_0)]} ,$$

for which from (2.32) we obtain

$$B(x) = -\frac{G_0(x, x_0)}{2xx_0} - \frac{8}{3\pi} P \int_0^\infty y^2 dy \frac{G_0(x, y)\left[\gamma - i\frac{\sqrt{3}}{2}\sqrt{1 - y^2}\right]}{2xy(y^2 - x_0^2)} B(y) , \tag{2.33}$$

where P is the principal value of the integral.

The function $G_0(x, y)$ has the form

$$G_0(x, y) = G^{(1)}(x, y) + i\pi G^{(2)}(x, y) \, ;$$

where

$$G^{(1)}(x, y) = \ln \left| \frac{x^2 + y^2 + xy - 3/4}{x^2 + y^2 - xy - 3/4} \right| \, ;$$

$$G^{(2)}(x, y) = \theta \left(\frac{x^2 + y^2 - 3/4}{xy} \right) \, ;$$

(2.34)

$$\theta(t) = \begin{cases} 1, |t| & < 1 \\ 0, |t| & > 1 \end{cases}$$

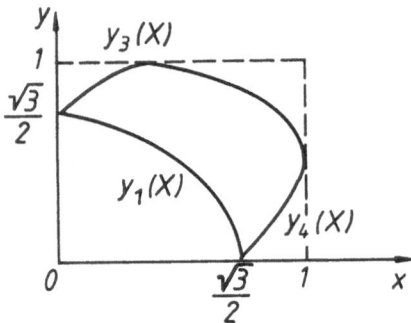

Fig. 2.5. The position of the singularities of the kernel of (2.34) of the integral (2.33)

As follows from (2.34), the kernel of (2.33) does indeed contain moving logarithmic singularities determined by

$$\left. \begin{array}{l} y_1 = -\frac{x}{2} - \frac{\sqrt{3}}{2}\sqrt{1 - x^2} \\ y_3 = \frac{x}{2} + \frac{\sqrt{3}}{2}\sqrt{1 - x^2} \end{array} \right\} 0 \le x \le \frac{\sqrt{3}}{2} \, ;$$

$$\left. \begin{array}{l} y_4 = \frac{x}{2} - \frac{\sqrt{3}}{2}\sqrt{1 - x^2} \\ y_3 = \frac{x}{2} + \frac{\sqrt{3}}{2}\sqrt{1 - x^2} \end{array} \right\} \frac{\sqrt{3}}{2} \le x < 1 \, .$$

The position of the singularities of the kernel of (2.34) in the (xy) plane is shown in Fig. 2.5. A review of the methods for solving integral equations with a singularity such as in (2.34) may be found in [32]. Here we shall briefly present a method based on the assumption that $B(x)$ is smooth. It is natural to assume that $B(x)$ is smooth within each of the two intervals $[0, 1]$ and $[1, \infty]$ [at $x = 1$, $B(x)$ has a discontinuous derivative] and allows an approximation on these intervals by the Lagrange interpolation polynomial [33]. After this approximation is made, the integrals containing singular terms can be computed analytically and the problem of finding $B(x)$ reduces to the solution of a set of algebraic equations for the interpolation coefficients. In practice, the procedure is the following: We represent the integral in the right-hand side of (2.33) as the sum of two terms:

$$P \int_0^\infty K(x,y)B(y)dy = \int_0^1 K(x,y)B(y)dy + P \int_1^\infty K(x,y)B(y)dy , \qquad (2.35)$$

where

$$K(x,y) = y^2 \frac{G_0(x,y)}{2xy} \frac{\gamma - i\sqrt{3/2}\sqrt{1-y^2}}{y^2 - x_0^2} . \qquad (2.36)$$

From (2.36) it follows that in the right-hand side of (2.35) the integrand in the first term contains moving logarithmic singularities, while in the second term only a fixed pole singularity occurs. Let us explicitly single out the singular factor $G_0(x,y)$ in the first integral in (2.35) and represent it as a sum of integrals over the intervals $h = 1/N$:

$$\int_0^1 dy\, G_0(x,y)\Phi(y) = \sum_{j=1}^N \int_{(j-1)h}^{jh} dy\, G_0(x,y)\Phi(y) , \qquad (2.37)$$

where

$$\Phi(y) = \frac{y^2}{2xy} \frac{\gamma - i\sqrt{3/2}\sqrt{1-y^2}}{y^2 - x_0^2} B(y) . \qquad (2.38)$$

Under this assumption the functions $B(y)$ and $\Phi(y)$ are smooth on the open interval $(0,1)$. Therefore, on each of the intervals $((j-1)h, jh)$ they can, to a sufficient degree of accuracy, be approximated by the Lagrange interpolation polynomial $L_k^{(j)}(y)$:

$$\Phi_j(y) = \sum_k L_k^{(j)}(y)\Phi(y_k) . \qquad (2.39)$$

The second integral in the right-hand side of (2.35) is computed by conventional methods.

In Fig. 2.6 the results of the solution of (2.33) for $E = 6.7\,\text{MeV}$ are displayed. As can be seen from this picture, the assumption that $B(x)$ is smooth on the intervals $(0,1)$ and $(1, \infty)$ is justified, therefore the procedure seems to be quite reliable.

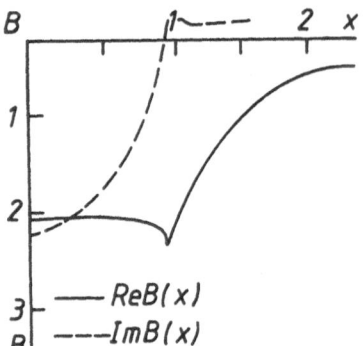

ReB(x)

———ImB(x)

Fig. 2.6. Solution of (2.23), $E = 6.7\,\mathrm{MeV}$ (—) $\mathrm{Re}\{B(x)\}$; (- - -)$\mathrm{Im}\{B(x)\}$

2.5 The Alt–Grassberger–Sandhas Form of the Faddeev Equations

We shall now consider the Faddeev equations in the form proposed by Alt, Grass-berger, Sandhas (AGS) [34a]. (Khelashvili proposed analogous equations, but his work has remained unpublished [34b].) These are the equations for the matrix elements, which represent physical amplitudes of transitions for the asymptotic states. (A less symmetrical choice has been proposed in [35].)

For convenience, we recall certain simple relations occurring for a two-particle system. Let there be a two-particle Hamiltonian $H = H_0 + V$ with corresponding Green functions $G(Z) = (Z - H)^{-1}$; $G_0(Z) = (Z - H_0)^{-1}$. As before, we introduce the t-matrix by the equality

$$G(Z) = G_0(Z) + G_0(Z)T(Z)G_0(Z) . \tag{2.40}$$

Then for the S-matrix in the momentum representation we have, as usual,

$$S_{pp'} = \delta(p - p') - 2\pi i\delta(E' - E)\langle p|T(E + i0)|p'\rangle . \tag{2.41}$$

Thus, for finding the t-matrix it is necessary to solve either

$$G(Z) = G_0(Z) + G_0(Z)VG(Z) , \tag{2.42}$$

for the Green function or the Lippmann–Schwinger equation

$$T(Z) = V + VG_0(Z)T(Z) . \tag{2.43}$$

In principle, both equations are equivalent, however, the t-matrix is less singular than the Green function; this is seen if one writes (2.40) in the momentum representation:

$$\langle p|G(Z)|p'\rangle = \frac{\delta(p - p')}{Z - p^2/2\mu} + \frac{1}{Z - p^2/2\mu}\langle p|T(Z)|p'\rangle\frac{1}{Z - p'^2/2\mu} . \tag{2.44}$$

Consider now the Hamiltonian of a three-particle system:

$$H = H_0 + \sum_\gamma V_\gamma, V_\gamma = V_{ij} , \quad \text{if } \gamma \neq i, j ; \quad \gamma, i, j = 1, 2, 3 .$$

Unlike the two-particle case, the three-particle Hamiltonian can be split into the channel Hamiltonian H_α and the "interaction"

$$H = H_\alpha + \bar{V}_\alpha \tag{2.45}$$

$$\bar{V}_\alpha = \sum_{\gamma \neq \alpha} V_\gamma . \tag{2.46}$$

From (2.45, 46), the Hamiltonian H_α describes the asymptotic case when the particle of index α is noninteracting, while the other two are in a bound state of potential V. Thus, the eigenstates $|\Phi_{\alpha n}\rangle$ of the Hamiltonian H_α, i.e., the channel eigenstates, are the products

$$|\Phi_{\alpha n}\rangle = |\Psi_{\alpha n}\rangle |q_\alpha\rangle , \tag{2.47}$$

where $|\Psi_{\alpha n}\rangle$ is the eigenfunction of the n-th bound state of the pair, while $|q\rangle$ is a plane wave describing the relative motion of the third particle and the center of mass of the bound pair.

Instead of the free Green function $G_0(Z)$, we introduce the channel Green function $G(Z) = (Z - H_\alpha)^{-1}$ and, by analogy with (2.40), we introduce the transition operator $U_{\beta\alpha}$:

$$G(Z) = \delta_{\alpha\beta} G_\alpha(Z) + G_\beta(Z) U_{\beta\alpha}(Z) G_\alpha(Z) . \tag{2.48}$$

We are compelled to introduce the channel Green functions $G_\alpha(Z)$, instead of the free Green functions $G_0(z)$, since now at infinity the channel solutions (2.47) exist, but not plane waves as in the two-body problem. By analogy with the definition of the two-particle S-matrix (2.41) we also define the matrix elements of the three-particle S-matrix:

$$S_{\beta m, \alpha n}(q'_\beta, q_\alpha) = \delta_{\beta\alpha} \delta_{mn} \delta(q'_\beta - q_\alpha)$$
$$-2\pi i \delta(E'_{\beta m} - E_{\alpha n}) \langle \Phi_{\beta m} | U_{\beta\alpha}(E_{\alpha n} + i\varepsilon) | \Phi_{\alpha n}\rangle . \tag{2.49}$$

By using the channel Green functions, instead of (2.42), for the three-particle Green function we obtain

$$G(Z) = G_\beta(Z) + G_\beta(Z) \sum_{\gamma \neq \beta} V_\gamma G(Z) . \tag{2.50}$$

Substituting into (2.50) the definition of the transition operator (2.48) we obtain the set of AGS equations for the operators $U_{\beta\alpha}$:

$$U_{\beta\alpha}(Z) = (1 - \delta_{\beta\alpha}) G_0^{-1}(Z) + \sum_{\gamma \neq \beta} T_\gamma(Z) G_0(Z) U_{\gamma\alpha} . \tag{2.51}$$

Comparison of (2.18, 51) reveals the absence of a diagonal term in the right-hand sides of both of these equations, i.e. of $U_{\beta\alpha}$ in (2.51) and $M_{\alpha\beta}$ in (2.18).

Thus, the physical meaning of the requirement that they be Fredholm-like, which was applied in obtaining (2.18), consists of taking into account the asymptotic states of a three-particle system, which leads to the Fredholm equations (2.51).

We shall now establish the relationship between the transition operators $U_{\alpha\beta}$ and the Faddeev operators $M_{\alpha\beta}$ [34c]. We rewrite the set of equations (2.51) as Lippmann–Schwinger equations

$$T = V + V \cdot \mathcal{G}_0 T \ . \tag{2.52}$$

where

$$T_{\beta\alpha} \equiv U_{\beta\alpha} \ ; \tag{2.53}$$

$$V_{\beta\alpha} \equiv (1 - \delta_{\alpha\beta})G_0^{-1} \ ; \tag{2.54}$$

$$\mathcal{G}_{0,\beta\alpha} \equiv \delta_{\beta\alpha} G_0 T_\alpha G_0 \ . \tag{2.55}$$

By analogy with the two-particle case we introduce together with the transition operator T, the Green operator $\mathcal{G}$:

$$\mathcal{G} = \mathcal{G}_0 + \mathcal{G}_0 T \mathcal{G}_0 \ . \tag{2.56}$$

It is easy to verify that $\mathcal{G}$ satisfies the equation

$$\mathcal{G} = \mathcal{G}_0 + \mathcal{G}_0 V \mathcal{G} \ . \tag{2.57}$$

Introducing (2.53, 55) into (2.56) and using (2.46) for $\mathcal{G}$ we obtain

$$\mathcal{G}_{\beta\alpha} = G_0 \left\{ \delta_{\beta\alpha} V_\alpha + V_\beta G V_\alpha \right\} G_0 \equiv G_0 M_{\beta\alpha} G_0 \ , \tag{2.58}$$

i.e., the elements of the Green function $\mathcal{G}$ are just the operators $M_{\alpha\beta}$ introduced by Faddeev (2.18). Substituting the definitions of $\mathcal{G}_0$ and V into (2.57) and using (2.58) we obtain the already familiar Faddeev equation for the operators $M_{\beta\alpha}$:

$$M_{\beta\alpha} = \delta_{\beta\alpha} T_\alpha + T_\beta G_0 \sum_{\gamma=\beta} M_{\gamma\alpha} \ .$$

Equations (2.51), just like the Faddeev equations, can be reduced to a one-dimensional equation if one uses the separable expansions for the pair t-matrix. Indeed, let us represent the pair operator T_γ as a sum of two terms:

$$T_\gamma = T_\gamma^s + T_\gamma' \ , \quad \gamma = 1, 2, 3 \ , \tag{2.59}$$

where

$$T_\gamma^s(Z) = \sum_{r=1}^{N_\gamma} |\gamma r\rangle \tau_{\gamma r}(Z) \langle \gamma r| \ , \tag{2.59a}$$

and T_γ' is a non-separable term, which in certain cases may be considered to be a perturbation. Substituting (2.59) into (2.51) we obtain a matrix equation which formally resembles the Lippmann–Schwinger equation:

$$T = V + V G_0 T \,, \tag{2.60}$$

where the following notation is used:

$$T_{\beta n, \alpha m} = \langle \beta n | G_0 U_{\beta \alpha} G_0 | \alpha m \rangle \,; \tag{2.61}$$

$$V_{\beta n, \alpha m} = \langle \beta n | G_0 U'_{\beta \alpha} G_0 | \alpha m \rangle \,; \tag{2.62}$$

$$G_{0, \beta n, \alpha m} = \delta_{\beta \alpha} \tau_{\alpha, nm} \,. \tag{2.63}$$

The operator $U'_{\beta \alpha}$ satisfies (2.51) in which T'_γ is used instead of T_γ; the index n assumes values from 1 to N_β in accordance with the number of separable terms included in the t-matrix like $1 \le m \le N$.

The matrix notation in (2.47,49) coincides with the one in (2.61–64) if the individual separable terms in (2.59a) correspond to real bound or resonance states in the two-particle subsystem. In the general case, the indices n and m may obviously assume values from a set larger than the number of bound two-particle states.

An important property of the operators (2.61–63) is that they do not act in the six-dimensional space of three-particle states, but in the three-dimensional space of states $|q_\alpha\rangle$, describing, as already noted, the free motion of the third particle α relative to the center of mass of the two-particle cluster $|\alpha m\rangle$. For example, the transition amplitude from the state α to the state β has the form

$$T_{\beta \alpha}^{nm}(q_\beta, q_\alpha) = \langle q_\beta | T_{\beta n, \alpha m} | q_\alpha \rangle \,. \tag{2.64}$$

In accordance with the above (2.60) transforms, after the separation of the angular variables, into a set of one-dimensional equations (naturally, under the condition that the operators $U'_{\beta \alpha}$ will be determined independently by means of an approximate method). Note that this reduction scheme can be implemented when the initial and/or final state contains three free particles.

2.6 The Karlsson and Zeiger Form of the Faddeev Equations

Until now, we have not been explicitly interested in the amplitudes describing the breakup processes, i.e., in the transition from two particles to three particles, for instance, n+d→n+n+p. A version of the Faddeev equations useful for dealing with such processes was proposed by *Karlsson* and *Zeiger* [36a]. Almost the same formulation was given earlier by *Grassberger* and *Sandhas* [36b].

We shall first derive the integral representation for the Faddeev component of the wave function $|\psi_\beta\rangle$, in which the singularities related to the existence of elastic scattering, rearrangement, and breakup are singled out explicitly. Then we shall determine the amplitudes corresponding to these processes as the residues at the respective poles. The equations for these residues are the *Karlsson–Zeiger equations*.

Let us introduce the complete set of functions for the channel Hamiltonian $H_\beta = H_0 + V_\beta$, $\beta = 1, 2, 3$:

$$\left\{ |\boldsymbol{p}_\beta \Phi_\kappa^\beta\rangle, \quad |\boldsymbol{p}_\beta \psi_{q_\beta}^{(-)}\rangle \right\} . \tag{2.65}$$

where $|\Phi_\kappa^\beta\rangle$ represents the wave function of the bound pair state β (we assume only to have a single bound state in each pair) of energy κ; $|\psi_{q_\beta}^{(-)}\rangle$ is the corresponding function from the continuous spectrum.

The Faddeev equations for the component of the wave function are

$$|\psi_\beta^{(+)}\rangle = \delta_{\beta\alpha}|\boldsymbol{p}_\alpha^{(0)}\Phi_\kappa^\alpha\rangle - G_0 B(E + i\varepsilon)t_\beta(E + i\varepsilon) \sum_{\gamma \neq \beta} |\psi_\gamma^{(+)}\rangle , \tag{2.66}$$

i.e., in the initial state there exists a bound pair in the channel α, $E = p^{2(0)}_\alpha 2\mu_\alpha - \kappa_\alpha^2$.

Let us now take advantage of the spectral representation for the channel Green function

$$G_\beta = \int \frac{|\boldsymbol{p}_\beta' \Phi_\kappa^\beta\rangle dp_\beta' \langle \boldsymbol{p}_\beta' \Phi_\kappa^\beta|}{p_\beta'^2/(2\mu_\beta) - \kappa^2 - Z}$$

$$+ \int \frac{|\boldsymbol{p}_\beta' \psi_{q_\beta}^{\beta(-)}\rangle dp_\beta dq_\beta \langle \boldsymbol{p}_\beta' \psi_{q_\beta'}^{\beta(-)}|}{p_\beta'^2/(2n_\beta) + q'^2/(2\mu_\beta) - Z} , \tag{2.67}$$

where $n_\beta = m_\beta(m_\alpha + m_\gamma)/(m_\alpha + m_\gamma + m_\beta)$, $\mu_\beta = m_\alpha m_\gamma/(m_\alpha + m_\gamma)$.

Applying (2.67) and the relation $G_0 t_\beta = G_\beta V_\beta$ for the projections $|\psi_\beta\rangle$ onto the channel components we obtain from (2.66)

$$\langle \boldsymbol{p}_\beta \Phi_\kappa^\beta | \psi_\beta^{(+)}\rangle = \delta_{\beta\alpha}\delta(\boldsymbol{p}_\alpha - \boldsymbol{p}_\alpha^0) - \frac{\mathcal{H}_{\beta\alpha}(\boldsymbol{p}_\beta; p_\alpha^0; E + (i\varepsilon)}{p_\beta^2/(2n_\beta) - \kappa_\beta^2 - E - i\varepsilon} ; \tag{2.68}$$

$$\langle \boldsymbol{p}_\beta \psi_{q_\beta}^{(-)} | \psi_\beta^{(+)}\rangle = \frac{\mathcal{E}_{\beta\alpha}(\boldsymbol{p}_\beta, \boldsymbol{q}_\beta; p_\alpha^{(0)}; E + i\varepsilon)}{p_\beta^2/(2n_\beta) + q_\beta^2/(2\mu_\beta) - E - i\varepsilon} , \tag{2.69}$$

where

$$\mathcal{H}_{\beta\alpha}(\boldsymbol{p}_\beta, \boldsymbol{p}_\alpha^{(0)} ; E + i\varepsilon) \equiv \langle \boldsymbol{p}_\beta \Phi_\kappa^\beta | V_\beta \sum_{\gamma \neq \beta} |\psi_\gamma^{(+)}\rangle ; \tag{2.70}$$

$$\mathcal{E}_{\beta\alpha}(\boldsymbol{p}_\beta, \boldsymbol{q}_\beta; \boldsymbol{p}_\alpha^{(0)}; E + i\varepsilon) \equiv \langle \boldsymbol{p}_\beta \psi_{q_\beta}^{(-)} | V_\beta \sum_{\gamma \neq \beta} |\psi_\gamma^{(+)}\rangle . \tag{2.71}$$

Using (2.68, 69) we obtain the desired integral representation for the Faddeev component $|\psi_\beta^{(+)}\rangle$ in the plane wave basis [35c]:

$$\langle \boldsymbol{p}\boldsymbol{q}_\beta | \psi_\beta^{(+)}\rangle = \delta_{\beta\alpha}\delta\left(\boldsymbol{p}_\alpha - \boldsymbol{p}_\alpha^{(0)}\right)\Phi_\kappa^\alpha(\boldsymbol{q}_\alpha) - \frac{\Phi_\kappa^\beta(\boldsymbol{q}_\beta)\mathcal{H}_{\beta\alpha}(\boldsymbol{p}_\beta, \boldsymbol{p}_\alpha^{(0)}; E + i\varepsilon)}{p_\beta^2/(2n_\beta) - \kappa_\beta^2 - E - i\varepsilon}$$

$$- \int dq_\beta' \frac{\psi_{q_\beta'}^{(-)}(\boldsymbol{q}_\beta)\mathcal{E}_{\beta\alpha}(\boldsymbol{q}_\beta, \boldsymbol{q}_\beta'; \boldsymbol{p}_\alpha^{(0)}; E + i\varepsilon)}{p_\beta^2/(2n_\beta) + q_\beta'^2/(2\mu_\beta) - E - i\varepsilon} . \tag{2.72}$$

From (2.72) it is clear that if the amplitudes are defined as residues of the wave function at the corresponding poles, then the functions $\mathcal{H}_{\beta\alpha}(\boldsymbol{p}_\beta, \boldsymbol{p}_\alpha^{(0)}; E = p_\beta^2/(2n\beta) - \kappa_\beta^2)$ represent the amplitudes for elastic scattering ($\alpha = \beta$) and rearrangement ($\alpha \neq \beta$). Similarly the expression for the breakup amplitude is

$$\sum_\beta \mathcal{E}_{\beta\alpha}\left[\boldsymbol{p}_\beta, \boldsymbol{q}_\beta; \boldsymbol{p}_\alpha^{(0)}; E = p_\beta^2/(2n_\beta) + q_\beta^2/(2\mu_\beta)\right] \ .$$

Representation (2.72) is essentially the three-particle analogue of the formula

$$\psi_{\boldsymbol{k}}^{(+)}(\boldsymbol{p}) = \delta(\boldsymbol{p} - \boldsymbol{k}) - \frac{t(\boldsymbol{p}, \boldsymbol{k}, k^2 + i\varepsilon)}{p^2/(2\mu) - k^2/(2\mu) - i\varepsilon} \ ,$$

already familiar to us from the two-particle problem. The equations for the half-mass amplitudes $\mathcal{H}_{\alpha\beta}$ and $\mathcal{E}_{\alpha\beta}$ are readily obtained, if representation (2.72) for the Faddeev component is substituted into the definitions (2.70,71). Below we shall be intersested in the equations for the off-shell amplitudes $\mathcal{H}_{\alpha\beta}$ and $\mathcal{E}_{\alpha\beta}$. However, before proceeding to obtain them, we shall establish the relationship between the amplitudes $\mathcal{H}_{\alpha\beta}, \mathcal{E}_{\alpha\beta}$ and the AGS amplitudes $U_{\alpha\beta}$.

It may be shown that there exists an analogue of the Möller operator which transforms the channel function into the exact Faddeev component,

$$|\psi_\beta^{(+)}\rangle = [\delta_{\beta\alpha} - G_0(E + i\varepsilon)K_{\beta\alpha}(E + i\varepsilon)]|\boldsymbol{p}_\alpha^{(0)}\Phi_\kappa^\alpha\rangle \ , \tag{2.73}$$

where $E = p_\alpha^{2(0)}/(2n_\alpha) - \kappa_\alpha^2$.

On account of the relationship between the operators $K_{\beta\alpha}$ and $U_{\beta\alpha}$

$$G_0 K_{\beta\alpha} = G_\beta V_\beta G_0 U_{\beta\alpha} = -G_0 t_\beta G_0 U_{\beta\alpha} \tag{2.74}$$

and of (2.72), and upon projecting (2.73) onto the channel states (2.65), we obtain the relationship between the amplitudes $\mathcal{H}_{\alpha\beta}$ and $\mathcal{E}_{\beta\alpha}$ and the matrix elements of the transition operator $U_{\beta\alpha}$:

$$\mathcal{H}_{\beta\alpha}(\boldsymbol{p}_\beta \boldsymbol{p}_\alpha^{(0)} \ ; \ E + i\varepsilon) = -\langle \boldsymbol{p}_\beta \Phi_\kappa^\beta | V_\beta G_0(E + i\varepsilon) U_{\beta\alpha}(E + i\varepsilon) | \boldsymbol{p}_\alpha^{(0)} \Phi_\kappa^\alpha \rangle \ . \tag{2.75}$$

$$\mathcal{E}_{\beta\alpha}(\boldsymbol{p}_\beta, \boldsymbol{q}; \boldsymbol{p}_\alpha^{(0)} \ ; \ E + i\varepsilon) = -\langle \boldsymbol{p}_\beta \psi_{\boldsymbol{q}}^{(-)} | V G_0(E + i\varepsilon) U_{\beta\alpha}(E + i\varepsilon) | \boldsymbol{p}_\alpha^{(0)} \Phi_\kappa^\alpha \rangle . \tag{2.76}$$

The most simple way to verify this is to compare (2.73) with the Faddeev equation (2.66). From the resulting equation for the operator $K_{\beta\alpha}$ and (2.51) (2.74) follows.

Obviously, on the mass shell

$$\langle \boldsymbol{p}_\beta \Phi_\kappa^\beta | V_\beta G_0(E + i0) = \langle \boldsymbol{p}_\beta \Phi_\kappa^\beta | \ . \tag{2.77}$$

Taking into account (2.77), from (2.75) we obtain

$$\mathcal{H}_{\beta\alpha}(\boldsymbol{p}_\beta, \boldsymbol{p}_\alpha^{(0)} \ ; \ E + i\varepsilon) = \langle \boldsymbol{p}_\beta \Phi_\kappa^\beta | U_{\beta\alpha} | \boldsymbol{p}_\alpha^{(0)} \Phi_\kappa^\alpha \rangle \ , \tag{2.78}$$

i.e., we once again verify, in terms of the operators $U_{\alpha\beta}$, that the amplitudes $\mathcal{H}_{\beta\alpha}$ describe elastic scattering and rearrangement. For $E \neq p_\alpha^{2(0)}/(2n_\alpha) - \kappa_\alpha^2$ and

$E \neq p_\beta^2/2n_\beta + q_\beta^2/(2\mu\beta)$, i.e., off the mass shell, (2.77) is no longer satisfied and for the off-shell continuation of the amplitudes (2.75, 76) we will have the expressions

$$\mathcal{H}_{\beta\alpha}\left(\boldsymbol{p}_\beta,\boldsymbol{p}_\alpha^{(0)};Z\right) = \langle\boldsymbol{p}_\beta\Phi_\kappa^\beta|V_\beta G_0(Z)U_{\beta\alpha}(Z)G_0(Z)V_\alpha|\boldsymbol{p}_\alpha^{(0)}\Phi_\kappa^\alpha\rangle\;; \qquad (2.79)$$

$$\mathcal{E}_{\beta\alpha}\left(\boldsymbol{p}_\beta,\boldsymbol{q}_\beta;\boldsymbol{p}_\alpha^{(0)},Z\right) = \langle\boldsymbol{p}_\beta\psi_{q_\beta}^{(-)}|V_\beta G_0(Z)U_{\beta\alpha}(Z)G_0(Z)V_\alpha|\boldsymbol{p}_\alpha^{(0)}\Phi_\kappa^\alpha\rangle\;. \qquad (2.80)$$

In averaging the equations for the off-shell amplitudes (2.79 80) we shall proceed from the AGS equations (2.51):

$$U_{\beta\alpha}(Z) = -\bar{\delta}_{\beta\alpha}G_0^{-1}(Z) - \sum_\gamma \bar{\delta}_{\beta\gamma}t_\gamma(Z)G_0(Z)U_{\gamma\alpha}(Z), \bar{\delta}_{\beta\alpha} \equiv 1-\delta_{\beta\alpha}.(2.80a)$$

Multiplying (2.80a) on the left by $V_\beta G_0$ and on the right by $G_0 V_\alpha$ and using (2.67, 79, 80) as well as the relation $G_0 t_\gamma = G_\gamma V_\gamma$ we obtain the desired equations:

$$\mathcal{H}_{\beta\alpha}\left(\boldsymbol{p}_\beta,\boldsymbol{p}_\alpha^{(0)};Z\right) = \mathcal{H}_{\beta\alpha}^{(0)}\left(\boldsymbol{p},\boldsymbol{p}_\alpha^{(0)};Z\right)$$
$$-\sum_{\gamma\neq\beta}\int d\boldsymbol{p}_\gamma' V_{\beta\gamma}^{\mathcal{H}\mathcal{H}}(\boldsymbol{p}_\beta,\boldsymbol{p}_\gamma')\frac{\mathcal{H}_{\gamma\alpha}\left(\boldsymbol{p}_\gamma'\boldsymbol{p}_\alpha^{(0)};Z\right)}{p_\gamma^2/(2n_\gamma)-\kappa_\gamma^2-Z}$$
$$-\sum_{\gamma=\beta}\int d\boldsymbol{p}_\gamma' d\boldsymbol{q}_\gamma' V_{\beta\gamma}^\varepsilon\left(\boldsymbol{p}_\beta,\boldsymbol{p}_\gamma',\boldsymbol{q}_\gamma'\right)\frac{\mathcal{E}_{\gamma\alpha}\left(\boldsymbol{p}_\gamma',\boldsymbol{q}_\gamma';\boldsymbol{p}_\alpha^{(0)};Z\right)}{p_\gamma^2/(2n_\gamma)+q_\gamma'^2/(2\mu_\gamma)-Z}\;; \qquad (2.81)$$

$$\mathcal{E}_{\beta\alpha}\left(\boldsymbol{p}_\beta,\boldsymbol{q}_\beta;\boldsymbol{p}_\alpha^{(0)};Z\right) = \mathcal{E}_{\beta\alpha}^{(0)}\left(\boldsymbol{p}_\beta,\boldsymbol{q}_\beta;\boldsymbol{p}_\alpha^{(0)};Z\right))$$
$$-\sum_{\gamma\neq\beta}\int d\boldsymbol{p}_\gamma' V_{\beta\gamma}^{\mathcal{E}\mathcal{H}}\left(\boldsymbol{p}_\beta,\boldsymbol{q}_\beta,\boldsymbol{p}_\gamma'\right)\frac{\mathcal{H}_{\gamma\alpha}\left(\boldsymbol{p}_\gamma',\boldsymbol{p}_\alpha^{(0)};Z\right)}{p_\gamma'^2/2n_\gamma-\kappa_\gamma^2-Z}$$
$$-\sum_{\gamma\neq\beta}\int d\boldsymbol{p}_\gamma' d\boldsymbol{q}_\gamma' V_{\beta\gamma}^{\mathcal{E}\mathcal{E}}\left(\boldsymbol{p}_\beta,\boldsymbol{q}_\beta;\boldsymbol{p}_\gamma',\boldsymbol{q}_\gamma'\right)\frac{\mathcal{E}_{\gamma\alpha}\left(\boldsymbol{p}_\gamma',\boldsymbol{q}_\gamma';\boldsymbol{p}_\alpha'^{(0)};Z\right)}{p_\gamma'^2/(2n_\gamma)+q_\gamma'^2/(2\mu_\gamma)-Z}\;,(2.82)$$

with

$$\mathcal{H}_{\beta\alpha}^{(0)}\left(\boldsymbol{p},\boldsymbol{p}^{(0)}\alpha;Z\right) \equiv \langle\boldsymbol{p}_\beta\phi_\kappa^\beta|V_\beta\sigma_0 V_\alpha|\boldsymbol{p}_\alpha^{(0)}\phi_\kappa^\alpha\rangle\bar{\delta}_{\beta\alpha}$$

$$\mathcal{E}_{\beta\alpha}^0\left(\boldsymbol{p}_\beta,\boldsymbol{q}_\beta;\boldsymbol{p}_\alpha^{(0)};Z\right) \equiv -\bar{\delta}_{\beta\alpha}\langle\boldsymbol{p}_\beta\psi_{q_\beta}^{(-)}|V_\beta\sigma_0 V_\alpha|\boldsymbol{p}_\alpha^{(0)}\phi_\kappa^\alpha\rangle$$

The effective potentials $V_{\beta\alpha}$ are determined as follows:

$$V_{\beta\gamma}^{\mathcal{H}\mathcal{H}}\left(\boldsymbol{p}_\beta,\boldsymbol{p}_\gamma'\right) = \frac{\Phi_\mathcal{H}^\beta(q_\beta^{(1)})\Phi_\mathcal{H}^\gamma(q_\gamma^{(2)})}{\left(q_\gamma^{(2)}\right)^2/(2n_\gamma)+\kappa_\gamma^2}\;; \qquad (2.83)$$

$$V_{\beta\gamma}^{\mathcal{H}\mathcal{E}}(\boldsymbol{p}_\beta,\boldsymbol{p}',\boldsymbol{q}') = \Phi_\mathcal{H}^\beta\left(q_\beta^{(1)}\right)\psi_{q_\gamma}^{(-)}\left(q_\gamma^{(2)}\right)\;; \qquad (2.84)$$

$$V_{\beta\gamma}^{\mathcal{E}\mathcal{H}}(\boldsymbol{p}_\beta, \boldsymbol{q}_\beta, \boldsymbol{p}'_\gamma) = -t_\beta \left(\boldsymbol{q}_\beta, \boldsymbol{q}_\beta^{(1)}; \frac{q_\beta^2}{2\mu_\beta} + i\varepsilon \right) \frac{\Phi_{\mathcal{H}}^\gamma \left(\boldsymbol{q}_\gamma^{(2)} \right)}{\left(q_\gamma^{(2)} \right)^2 /(2\mu_\gamma) + \kappa_\gamma^2} . \qquad (2.85)$$

$$V_{\beta\gamma}^{\mathcal{E}\mathcal{E}}(\boldsymbol{p}_\beta, \boldsymbol{q}_\beta, \boldsymbol{p}'_\gamma, \boldsymbol{q}'_\gamma) = t_\beta \left(\boldsymbol{q}_\beta, \boldsymbol{q}_\beta^{(1)}; \frac{q_\beta^2}{2\mu_\beta} + i\varepsilon \right) \psi_{\boldsymbol{q}'_\gamma}^{(-)} \left(\boldsymbol{q}_\gamma^{(2)} \right) . \qquad (2.86)$$

Here the following notation is utilized:

$$\boldsymbol{q}_\beta^{(1)} = \frac{m_\gamma}{m_\alpha + m_\gamma} \boldsymbol{p}_\beta + \boldsymbol{p}'_\gamma ; \quad \bar{\Phi}_{\mathcal{H}}^\beta \boldsymbol{q}_\beta = - \left(\frac{q_\beta^2}{2\mu_\beta} + \kappa_\beta^2 \right) \phi_{\mathcal{H}}^\beta (\boldsymbol{q}_\beta) ;$$

$$\boldsymbol{q}^{(2)} = -\boldsymbol{p}_\beta - \frac{m_\beta}{m_\alpha + m_\beta} \boldsymbol{p}'_\gamma .$$

As can be seen from expressions (2.81–86), the kernels of the Faddeev equations in the Karlsson–Zeiger form are determined by effective potentials independent of the energy Z. In practice, this turns out to be an extremely useful property. Applying a simple transformation in the equations for partial amplitudes one can arrive at equations with real effective potentials, which is also extremely convenient. Finally, the input data for equations (2.81, 83) consist of pair wave functions of the discrete and continuous spectra, as well as of the half-mass-shell pair T-matrices.

2.7 Solution by Means of the Method of Moments and Spline Functions

Let us again return to the method of moments and apply it to solving the Faddeev equations. We shall consider two versions of the method, to be applied to the problem of S-scattering of a particle in a bound state where all masses are equal. The first version consists of constructing auxiliary functions by iterations (1.134) of one-dimensional Faddeev equations (for the quartet and doublet nd-scattering).

Let us have a one-dimensional equation such as (2.30) or (2.60) for the function $|\chi\rangle$:

$$|\chi\rangle = |Z_0\rangle + Z\Delta|\chi\rangle . \qquad (2.87)$$

Just as in (1.134), we construct an iteration chain of functions $|\chi_n\rangle$ which represent moments of the kernel $Z\Delta$:

$$|\chi_0\rangle = |Z_0\rangle ; \quad |\chi_n\rangle = |\chi_0\rangle + Z\Delta|\chi_{n-1}\rangle , \quad n = 1 \dots N . \qquad (2.88)$$

As usual, we search for the solution in the form of a linear combination of the function $|\chi_n\rangle$:

$$|\chi^N\rangle = \sum_{n=0}^{N} a_n^N |\chi_n\rangle . \qquad (2.89)$$

Substituting (2.89) into (2.87) and projecting onto the functions $\langle \chi_m | \Delta$, we obtain a set of equations for the coefficients a_n^N:

$$\sum_{n=0}^{N} \{ \langle \chi_m | \Delta | \chi_n \rangle - \langle \chi_m | \Delta Z \Delta | \chi_n \rangle \} a_n^N = \langle \chi_m | \Delta | Z_0 \rangle . \tag{2.90}$$

In the momentum representation (2.87) becomes

$$\chi(p, k) = Z_0(p, k, k) - \frac{2}{\pi} \int_0^\infty \frac{Z(p, p', k)}{p'^2 - k^2 - i\varepsilon} \chi(p', k) p'^2 dp' . \tag{2.91}$$

The solution of equation (2.91) is related to the three-particle scattering phase:

$$\chi(k, k) = -\frac{3}{4k} \exp[i\delta(k)] \sin \delta(k) , \tag{2.92}$$

while the functions Z_0 and Z are in the same form as the already familiar (2.31) angular integrals:

$$Z_0(p, p', k) = b^s \int_{-1}^{1} \frac{\Phi(q_1)\Phi(q_2)dx}{p^2 + p'^2 + pp'x - (3/4)k^2 - \varepsilon} , \tag{2.93}$$

where $b^s = -1$ for $s = 0$; $b^s = \frac{1}{2}$ for $s = \frac{1}{2}$; s is the spin of the particle; $q_1 = (\frac{1}{4}p^2 + p'^2 + pp'x)^{1/2}$; $q_2 = (p^2 + \frac{1}{4}p' + pp'x)^{1/2}$; ε is the binding energy of the pair on which scattering takes place; $\Phi(q) = c/(q^2 + \beta^2)$ is the form factor of the separable S-wave Yamaguchi potential; $c^2 = 2\beta\alpha(\beta + \alpha)^3$; $\alpha = \sqrt{\varepsilon}$; β represents the range of the Yamaguchi potential

$$Z(p, p', k) = \frac{3}{4} Z_0(p, p', k) S \left[\frac{4}{3}(k^2 - p'^2) - \varepsilon \right] \tag{2.94}$$

and

$$S(-\xi^2) = \frac{(\beta + \xi)^2(\alpha + \xi)}{\alpha(\alpha + \beta)(\alpha + \xi + 2\beta)} . \tag{2.95}$$

Numerical computations reveal [37] that for moments $0 \leq p \leq 2(\text{fm}^{-1})$ the series (2.89) converges to the exact value of the function [obtained by direct pointwise integration of (2.91)] for $N = 4$ in the case of doublet scattering and for $N = 2$ in the case of quartet scattering.

The second version of the method of moments [38] is based on another set of auxiliary functions $|\chi_n\rangle$, the so-called *spline functions* [39]. Let us define these functions. For convenience we transform the integration interval in (2.91) into the interval $[0, 1]$. On this interval we introduce an ordered set of points C_i, i.e., $0 = C_1 < C_2 \ldots < C_{N-1} < C_N = 1$. Then the basic spline function χ_m^i will be defined as

$$\chi_m^i(x) = [\theta(x - C_i) - \theta(x - C_{i+1})](x - C_i)^{m-1} , \tag{2.96}$$

where $\Theta(x) = \begin{cases} 1 & , x > 0 \\ 0 & , x < 0 \end{cases}$ and $1 \le i \le N$, $1 \le m \le s$. The quantity s fixes the number of independent basis functions for each point C_i. From the definition it is clear that the function $\chi^i_m(x)$ differs from zero only within the interval $[C_i, C_i + 1]$. We shall approximate a function $\Phi(x)$ given on the same interval by the sum

$$\Phi(x) = \sum_{i=1}^{N} \sum_{m=1}^{s} r^i_m \chi^i_m(x) . \tag{2.97}$$

The coefficients of this expansion r^i_m are determined from the condition $r^i_1 = \Phi(C_i)$ and from the requirement that $\Phi(x)$ and its first $(s-1)$ derivates should be continuous at all the internal points C_i of the interval $[0, 1]$. These conditions lead to a set of linear equations for r^i_m. If a uniform distribution of the points C_i is used, i.e., $C_{i+1} - C_i = \delta$, then one can obtain the following estimate for the uncertainty due to the approximation (2.97): $|\Phi(x) - \tilde{\Phi}(x)| < \text{const} \cdot \delta^s$.

From the definition of the expansion (2.97) it follows that the approximations in the spline functions (2.96) can be applied only to continuous functions. Therefore in (2.91), at energies not exceeding the breakup threshold, we shall approximate a smooth function $Z(p, p', k)$ by splines. In addition we shall take into account explicitly the asymptotic behavior of $Z(p, p', k)$ at large p. From (2.94) we find that for $p \gg p', k$, $Z(p, p', k) \sim p^{-6}$, while for $p \sim p' \gg k$, $Z(p, p', k) \sim p^{-4}$. In accordance with the above we introduce a new function

$$Y(p, p', k) = \frac{1 + p'^2}{(1 + p^2)^3} \tilde{Y}_1(p, p', k) , \tag{2.98}$$

which we shall approximate as

$$\tilde{Y}_1(p, p', k) = \sum_{i,j=1}^{N} \sum_{m,n=1}^{s} \beta^{ij}(k) \chi^i_m(p) \chi^j_m(p'); , \tag{2.99}$$

where

$$\chi^i_n(p) = [\theta(x - C_i) - \theta(x - C_{i+1})](x - C_i)^{n-1} ;$$
$$p = p_0 \frac{1 - x}{1 - x} ; \quad x \in [-1, 1] . \tag{2.100}$$

The scaling factor p_0 may be used to optimize the solution.

For convenience, instead of (2.91), we shall take another equation where only the principal value of the integral is taken into account. This is allowed by the K-matrix form of (2.91). Then, bearing in mind (2.99), we obtain for $K(p, k)$ the expansion

$$\tilde{K}(p, k) = Z(p, k, k) - (1 + p^2)^{-3} \sum_{i=1}^{N} \sum_{m=1}^{s} \alpha^i_m(k) \chi^i_m(p) , \tag{2.101}$$

with

$$K(k, k) = -\frac{3}{4} \frac{1}{k \cot \delta(k)} .$$

The coefficients $a_m^i(k)$ are found from the set of linear equations

$$A\alpha = B . \tag{2.102}$$

Here

$$A_{mn}^{ij} = \delta_{ij}\delta_{mn} + \sum_{l=1}^{s} \beta_{ml}^{ij} l_{ln}^{lj} ; \tag{2.103}$$

$$B_m^i = \sum_{j=1}^{N} \sum_{n=1}^{s} \beta_{mn}^{ij} g_n^i ; \tag{2.104}$$

$$l_{mn}^j(k) = \frac{2}{\pi} \int_{p_j}^{p_j+1} \frac{\chi_m^j(p)\chi_n^j(p)}{(1+p^2)^2(p^2-k^2)} p^2 dp ; \tag{2.105}$$

$$g_n^j(k) = \frac{2}{\pi} \int_{p_j}^{p_j+1} Z(p, k, k) \frac{(1+p^2)\chi_m^j(p)}{p^2-k^2} p^2 dp , \tag{2.106}$$

where

$$p_j = p_0(1 + C_j)/(1 - C_j) .$$

The parameters of the pair potential describing low energy np–interaction in the triplet state were computed to yield $\varepsilon = 0.054\,\mathrm{fm}^{-2}$, $\beta = 1.444\,\mathrm{fm}^{-1}$. The most stable solutions were obtained for $p_0 = 0.5\,\mathrm{fm}^{-1}$. This method yielded very good convergence for the half-mass shell quartet amplitude at all energies up to the breakup threshold of the deuteron. Table 2.1 shows the convergence of the results for the quartet scattering length using this method.

Table 2.1.

N	6	8	10	exact
$a_{3/2}^{[\mathrm{fm}]}$	6.298	6.286	6.284	6.284

2.8 Unitarization of Approximate Solutions

We shall now present some approximate methods for determining three-particle amplitudes based on the *unitarity condition*, i.e., we shall assume the approximate amplitude to satisfy the condition of two-particle or three-particle unitarity. We shall consider three different formulations. These have turned out to be extremely effective as approximate descriptions of real three-particle systems.

The first formulation is based on an approximate solution of the Faddeev equations and is suitable for local and nonlocal pair potentials [40]. In this version of unitarization the set of the Faddeev equations is rewritten in the form of two related sets of equations. As before, the second set remains multidimensional, while the kernel of the first one is transferred to the mass shell (the first set becomes one-dimensional, and in the case of a central interaction it even transforms into an algebraic set). Thus, if one finds (for instance, by iteration) some approximate solution of the second set, then the solution of the first set leads immediately to an explicit and unitary expression for the amplitude. Two-particle unitarity is intended, i.e., no three-particle states are present among the intermediate ones.

We shall now proceed with the explicit exposition of this formulation. Assume there to be a set of three particles of masses m_1, m_2, m_3 and pair potentials V_1, V_2, V_3, where, for example, $V_1 = V_{23}$ represents the interaction between particles 2 and 3. The total Hamiltonian of the system is

$$H = H_0 + \sum_{\alpha=1}^{3} V_\alpha .$$

We now introduce the following operators:

$$G(Z) = -(Z - H)^{-1} ; \tag{2.107}$$

$$G_\alpha(Z) = -(Z - H_\alpha)^{-1} ; \tag{2.108}$$

$$T_\alpha(Z) = V_\alpha - V_\alpha G_\alpha(Z) V_{\alpha'} , \quad \alpha = 1, 2, 3 ; \tag{2.109}$$

where $H_\alpha = H_0 + V_\alpha$ is the channel Hamiltonian. For the channel operators we require

$$G_0(Z) T_\alpha(Z) = G_\alpha(Z) V_\alpha ; \quad T_\alpha(Z) G_0(Z) = V_\alpha G_\alpha(Z) \tag{2.110}$$

(Chap. 1). Operators (2.108, 109) are defined in the space of three-particle states; we shall now introduce their analogues in the two-particle subspace:

$$g_\alpha(Z) = -(Z - h_\alpha)^{-1} ; \tag{2.111}$$

$$t_\alpha(Z) = V_\alpha - V_\alpha g_\alpha(Z) V_\alpha , \quad \alpha = 1, 2, 3 . \tag{2.112}$$

We now write (2.21) for the pair α:

$$\langle \boldsymbol{p}_\alpha \boldsymbol{q}'_\alpha | T_\alpha(Z) | \boldsymbol{p}_\alpha \boldsymbol{q}_\alpha \rangle = \delta(\boldsymbol{p}'_\alpha - \boldsymbol{p}_\alpha) \langle \boldsymbol{q}'_\alpha | t_\alpha(Z - p_\alpha^2/(2M_\alpha)) | \boldsymbol{q}_\alpha \rangle , \tag{2.113}$$

where as before, for example, $M_1 = m_1(m_2 + m_3)/(m_1 + m_2 + m_3)$. We now introduce the operators $R_{\beta\alpha}^\pm$ [35]:

$$R_{\beta\alpha}^+(Z) = \delta_{\beta\alpha} V_\alpha + W_{\beta\alpha}(Z) ; \tag{2.114}$$

$$R_{\beta\alpha}^-(Z) = \bar{\delta}_{\beta\alpha} V_\beta + W_{\beta\alpha}(Z) , \tag{2.115}$$

where the operators $W_{\beta\alpha}$ are defined through the complete Green function

$$W_{\beta\alpha}(Z) = \sum_{\gamma} \bar{\delta}_{\gamma\alpha}\bar{\delta}_{\gamma\beta}V_{\gamma} - \sum_{\gamma,\delta} \bar{\delta}_{\gamma,\beta}\bar{\delta}_{\delta\alpha}V_{\gamma}G(Z)V_{\delta} .$$ (2.116)

The operators $R^{\pm}_{\beta\alpha}$ have the property that their matrix elements with respect to channel functions $|\varphi_{\alpha m}\rangle$ represent physical transition amplitudes T from the channel α to the channel β, on the mass shell:

$$\langle \beta m|T|\alpha n\rangle = \langle \varphi_{\beta m}(E)|R^{\pm}_{\beta\alpha}(E + i\varepsilon)|\varphi_{\alpha n}(E)\rangle .$$ (2.117)

Thus, on the mass shell the matrix elements of the operators $R^{\pm}_{\beta\alpha}$ coincide with the previously introduced (2.48,49) matrix elements of the operators $U_{\beta\alpha}$. We also note that the operators $R^{\pm}_{\beta\alpha}$ have identical matrix elements on the mass shell. This is readily verified from their definition and the identity

$$\langle \varphi_{\beta m}(E)|V_{\beta}|\varphi_{\alpha n}(E)\rangle = \langle \varphi_{\beta m}(E)|V_{\alpha}|\varphi_{\alpha n}(E)\rangle .$$ (2.118)

The operators $R^{\pm}_{\beta\alpha}$ satisfy the Lovelace form of the Faddeev equations:

$$R^{+}_{\beta\alpha}(Z) = \sum_{\gamma} \bar{\delta}_{\gamma\beta}V_{\gamma} - \sum_{\gamma} \bar{\delta}_{\gamma\alpha}R^{+}_{\beta\gamma}(Z)G_0(Z)T_{\gamma}(Z) ;$$ (2.119)

$$R^{-}_{\beta\alpha}(Z) = \sum_{\gamma} \bar{\delta}_{\gamma\alpha}V_{\gamma} - \sum_{\gamma} \bar{\delta}_{\gamma\beta}T_{\gamma}(Z)G_0(Z)R^{-}_{\gamma\alpha}(Z) .$$ (2.120)

In accordance with (2.114,115), as well as (2.109,110), we transform to equations for the operators $W_{\beta\alpha}$:

$$W_{\beta\alpha} = \sum_{\gamma} \bar{\delta}_{\gamma\beta}\bar{\delta}_{\gamma\alpha}T_{\gamma} - \sum_{\gamma} \bar{\delta}_{\gamma\alpha}W_{\beta\gamma}G_0T_{\gamma} ;$$ (2.121)

$$W_{\beta\alpha} = \sum_{\gamma} \bar{\delta}_{\gamma\beta}T_{\gamma} - \sum_{\gamma} \bar{\delta}_{\gamma\beta}\bar{\delta}_{\gamma\beta}T_{\gamma}G_0W_{\gamma\alpha} .$$ (2.122)

We rewrite (2.122) in a form convenient for the approximations to be performed. To this end we represent the operator T_{γ} as a sum of two terms:

$$T_{\gamma} = T^{(1)}_{\gamma} + T^{(2)}_{\gamma} ,$$ (2.123)

the explicit form of which will be given below. We introduce the operator $W^{(1)}_{\beta\alpha}$ satisfying (2.121), in which $T^{(1)}_{\gamma}$ is substituted for T_{γ}:

$$W^{(1)}_{\beta\gamma} = \sum_{\gamma} \bar{\delta}_{\gamma\beta}\bar{\delta}_{\gamma\alpha}T^{(1)}_{\gamma} - \sum_{\gamma} W^{(1)}_{\beta\gamma}G_0T^{(1)}_{\gamma}\bar{\delta}_{\gamma\alpha} ;$$ (2.124)

as well as the operator $R^{(1)}_{\beta\alpha}$ corresponding to it:

$$R^{(1)}_{\beta\alpha} = \bar{\delta}_{\beta\alpha}V_{\alpha} + W^{(1)}_{\beta\alpha} .$$ (2.125)

By substituting (2.123) into (2.121) and, upon transition to the matrix form, inverting the part of the equation $\sim T^{(1)}$ making use of (2.124) it can be readily shown that the operator $W_{\beta\alpha}$ satisfies the equation

$$W_{\beta\alpha} = W_{\beta\alpha}^{(1)} + \sum_\gamma \left[\bar{\delta}_{\beta\gamma} - W_{\beta\gamma}^{(1)}G_0\right] T_\gamma^{(2)} \left[\bar{\delta}_{\gamma\alpha} - G_0 W_{\gamma\alpha}\right] . \tag{2.126}$$

This derivation is much more straightforward than that of [40]. We choose $T_\gamma^{(2)}$ in the form

$$T_\gamma^{(2)}(E + i\varepsilon) = V_\gamma G_\gamma'(E)V_\gamma' \tag{2.127}$$

where

$$G_\gamma'(E) = -i\pi\delta(E - H_\gamma)P_\gamma , \quad \gamma = 1,2,3 , \tag{2.128}$$

and P_γ is the projection operator onto the channel eigenstate $|\varphi_{\gamma n}\rangle$ (i.e., the eigenstate of the Hamiltonian H_γ).

Thus, we have achieved the separation of the Faddeev equations (2.121) into two sets (2.124, 126). Owing to the choice of $T_\gamma^{(2)}$ in the form of (2.127, 128), the kernels of equations (2.126) are transferred to the mass shell, i.e., (2.126) represent integral equations for the three-particle amplitudes taken on the mass shell.

Using the relations

$$G_\gamma'(E)V_\gamma G_0(E + i\varepsilon) = G_\gamma'(E) ;$$
$$G_0(E + i\varepsilon)V_\gamma G_\gamma'(E) = G_\gamma'(E) , \tag{2.129}$$

as well as (2.119, 120, 125), it is possible to simplify (2.126) and to transform to equations in $R_{\beta\alpha}$:

$$R_{\beta\alpha}(E + i\varepsilon) = R_{\beta\alpha}^{(1)}(E + i\varepsilon) + \sum_\gamma R_{\beta\gamma}^{(1)}(E + i\varepsilon)G_\gamma'(E)R_{\gamma\alpha}(E + i\varepsilon) . \tag{2.130}$$

Until now, these are exact equations. We shall perform an approximation for the operators $R_{\beta\alpha}^{(1)}$, namely, in (2.125) we shall only retain the zeroth iteration

$$R_{\beta\alpha}^{(1)} \approx \tilde{R}_{\beta\alpha}^{(1)} = \bar{\delta}_{\beta\alpha}V_\alpha + \sum_\gamma \bar{\delta}_{\gamma\beta}\bar{\delta}_{\gamma\alpha}T_\gamma^{(1)} . \tag{2.131}$$

Using (2.131) for the case of $\alpha = 1$ (i.e., particle 1 is incident on the bound state of particles 2 and 3) in the absence of bound states in the potentials V_{13} and V_{12}, we obtain from (2.130) a simple equation for the operator of an "elastic" transition:

$$R_{11} = (T_{13} + T_{12}) + (T_{13} + T_{12})G_1'R_{11} . \tag{2.132}$$

We shall solve (2.130) for a system of three spinless particles. We write (2.130) in the momentum representation:

$$\langle \beta k'_\beta |T|\alpha k_\alpha \rangle = \langle \beta k'_\beta |I|\alpha k_\alpha \rangle - i\pi \sum_{\delta=1}^{3} \int dk''_\delta \times \langle \beta k'_\beta |I|\delta k''_\delta \rangle$$

$$\times \; \delta(E - E''_\delta) \langle \delta k''_\delta |T|\alpha k_\alpha \rangle \; . \tag{2.133}$$

Here

$$\langle \beta k'_\beta |T|\alpha k_\alpha \rangle \equiv \langle \beta k'_\beta |R_{\beta\alpha}|\alpha k_\alpha \rangle \; ;$$

$$\langle \beta k'_\beta |I|\alpha k_\alpha \rangle \equiv \langle \beta k'_\beta |R_{\beta\alpha}^{(1)}|\alpha k_\alpha \rangle \; .$$

For identical particles in (2.133) only two independent amplitudes are present, the direct T^D and the exchange T^R:

$$\langle k'|T^D|k \rangle = \langle \alpha k'|T|\alpha k \rangle \; , \quad \alpha = 1, 2, 3 \; ; \tag{2.134}$$

$$\langle k'|T^R|k \rangle = \langle \beta k'|T|\alpha k \rangle \; , \quad \beta \neq \alpha \; . \tag{2.135}$$

Introducing the symmetrized (or antisymmetrized) amplitude

$$\langle k'|T|k \rangle = \langle k'|T^D|k \rangle + 2\langle k'|T^D|k \rangle \; , \tag{2.136}$$

we split up the set (2.133). For T we obtain

$$\langle k'|T|k \rangle = \langle k'|I|k \rangle - i\pi \int dk'' \langle k'|I|k'' \rangle \delta(E - E'') \langle k''|T|k \rangle \; , \tag{2.137}$$

where

$$\langle k'|l|k \rangle = \langle k'|I^D|k \rangle + 2\langle k'|l^R|k \rangle \; . \tag{2.138}$$

Separating out the angular variables in (2.137) we find for $T(k, k')$

$$T_l(k, k') = I_l(k, k')/[1 - i I_l(k, k')] \; . \tag{2.139}$$

In Fig. 2.7 the differential cross section for an incident particle energy of 14.1 MeV constructed by using the amplitudes of (2.139) is compared with the exact solution. (Naturally, in both cases a separable potential was used.) As can

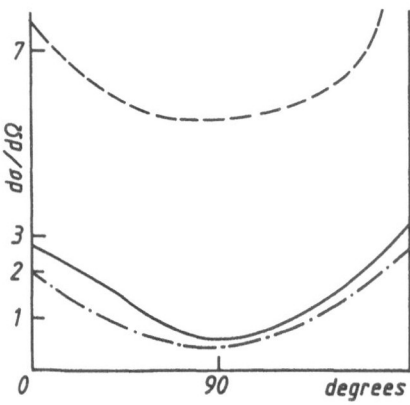

Fig. 2.7. Differential cross section for elastic scattering of a particle of energy $E_{kin} = 14.1$ MeV on a bound state of two particles: (——) exact solution; (·······) unitarized approximate solution (2.139); (- - -) impulse approximation. (The cross section is given in relative units, the particles are identical)

be seen from the figure, the amplitude given by (2.139) is extremely close to the exact one, i.e., the unitarization procedure realized by (2.130) leads to a rather fundamental renormalization of the "impulse approximation" I_l.

2.9 Dispersion Methods in the Three-Body Problem

The second method of utilizing the unitarity condition to find approximate three-particle amplitudes consists of the combined application of the unitarity condition and of the analytic properties of the amplitude as a function of energy [28, 41–43]. In this approach, equations for the amplitudes for physical processes are based on the application of the Riemann boundary-value problem [44]. We shall briefly recall its formulation. Let there be two functions on the real axis, $D(x)$ and $H(x)$. To be found are the two functions $F^{\pm}(z)$, analytic in the upper and the lower half-planes, respectively, the limit values of which on the real axis satisfy the boundary condition

$$F^+(x) = D(x)F^-(x) + H(x) .\qquad(2.140)$$

All the functions encountered in the Riemann problem $(F^{\pm}, D - 1, H)$ must be quadratically integrable and satisfy the Hölder boundary condition [44]. The simplest case of the Riemann boundary-value problem is the step-function problem, i.e., the case when (2.140) becomes

$$F^+(x) - F^-(x) = H(x) .\qquad(2.141)$$

The solution of this boundary-value problem are the functions

$$F^+(z) = \frac{1}{\sqrt{2\pi}} \int\limits_0^\infty h(t) \exp(itz)dt ,$$

$$F^-(z) = \frac{1}{\sqrt{2\pi}} \int\limits_{-\infty}^0 h(t) \exp(itz)dt ,$$

$$(2.142)$$

where

$$h(t) = \frac{2}{\sqrt{2\pi}} \int\limits_{-\infty}^\infty H(x) \exp(itz)dx .$$

Precisely such a situation is encountered in the quantum-mechanical two-body problem. Indeed, for the step in the partial amplitude $f_l(E)$ we have

$$\tilde{f}_l(E + i0) - \tilde{f}_l(E - i0) = a_l(E) , \quad -\infty < E \le -m^2/4 ,$$

$$\tilde{f}_l(E) \equiv \sqrt{E} f_l(E) ,$$

$$(2.143)$$

where the function $\alpha_l(E)$ can be considered as given. The parameter m signifies that scattering is considered only for potentials for which the range r_0 does not exceed m^{-1}.

The scattering amplitude is expressed through the Jost function $F_l(k)$ as

$$\frac{f_l(E)}{\sqrt{E}} = \frac{[F_l(k) - F_l(-k)]/2ik}{F_l(-k)} = \frac{N_l(E)}{D_l(E)}, \quad E = k^2. \tag{2.144}$$

From the properties of the Jost function it is known that $D(E)$ possesses only a right-hand cut from $E = 0$ to ∞. In addition $D_l(E) \to 1$ as $|E| \to \infty$. On the other hand, $N_l(E)$ has no right-hand cut since it depends only on k^2; it does, however, have a left-hand cut from $-\infty$ to $-m^2/4$. Both the functions $F_l(\pm k)$ tend towards unity as $|k| \to \infty$, therefore $N_l(E) \to 0$ as $|E| \to \infty$. The Jost function $F_l(k)$ exhibits the property: $F_l(k) = F_l^*(-k^*)$, from which we immediately obtain that $D_l(E)$ is real for $E < 0$, while $N_l(E)$ is real in the region of $E > 0$. Taking into account these properties and (2.143, 144) we obtain

$$\frac{N_l(E + i0) - N_l(E - i0)}{2iD_L(E)} = \frac{\operatorname{Im} N_l(E)}{D_l(E)} = \frac{-\alpha_l(E)}{2|E|^{1/2}};$$
$$-\infty < E \le -\frac{m^2}{4} \tag{2.145}$$

or for the step in N_l:

$$N_l(E + i0) - N_l(E - i0) = \frac{i\alpha_l(E)}{2|E|^{1/2}} D_l(E).$$

Further, from (2.144) and the unitarity condition $\operatorname{Im} \tilde{f}_l(k) = |\tilde{f}_l(k)|^2$ we obtain

$$\operatorname{Im} \tilde{f}_l^{-1}(E) = -\operatorname{Im} \tilde{f}_l |\tilde{f}_l|^{-2} = -1 = \operatorname{Im} D_l(E)/\sqrt{E} N_l(E)$$
$$= [D_l(E + i0) - D_l(E - i0)]/2i\sqrt{E} N_l(E).$$

Thus, for a step in $D_l(E)$ we have

$$D_l(E + i0) - D_l(E - i0) = -2i\sqrt{E} \theta(E) N_l(E). \tag{2.146}$$

As a result, we ultimately obtain a set of integral equations for $N_l(E)$ and $D_l(E)$ [known in physical literature as the N/D-equations (N/D-method)]:

$$N_l(E) = -\frac{1}{2\pi} \int_{-\infty}^{-m^2/4} \frac{\alpha_l(E')D_l(E')}{\sqrt{E'}(E' - E)} dE'; \tag{2.147}$$

$$D_l(E) = 1 - \frac{1}{\pi} \int_0^\infty \frac{\sqrt{E'} N_l(E')}{E' - E} dE' \tag{2.148}$$

or, which is the same, an equation for $D_l(-E)$:

$$D_l(-E) = 1 - \frac{1}{2\pi} \int\limits_{m^2/4}^{\infty} \frac{\alpha_l(-E')D_l(-E')}{\sqrt{E'}(\sqrt{E'} + \sqrt{E})} dE \ . \tag{2.149}$$

Thus, fixing the step in $\alpha_l(E)$ and solving (2.147–149) one can fully determine the scattering amplitude on the mass shell as a function of energy. Thus the information on the interaction dynamics contained in the potential (in the approach based on the Schrödinger equation), is in this approach given with the aid of some approximate function $\alpha_l(E)$.

As a first example of the application of (2.147–149) to three-particle systems, we shall consider a system consisting of a neutron and a deuteron in the quartet state. We shall be interested in finding the quartet nd-scattering length.

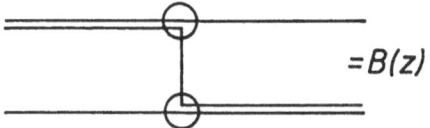

$= B(z)$

Fig. 2.8. Diagram of nd-scattering corresponding to the singularity nearest to the physical region

By analogy with (2.144), we represent the S-wave of the nd-scattering amplitude as

$$f(E) = \frac{\exp[i\delta(k)]}{k} \sin\delta(k) = \frac{N(E)}{D(E)} \ . \tag{2.150}$$

Just as in the previous case, the amplitude given in (2.150) has a unitary right-hand cut in the energy and a "dynamic" left-hand cut on a finite energy interval to be defined below. If the contribution of the virtual breakup of the deuteron is not taken into account, then, obviously, the unitarity condition will be of a two-particle form: $\mathrm{Im} f^{-1}(E) = -k$. The step in the amplitude on the left-hand cut $\alpha(E)$, as in the previous case, will be given on the basis of some additional arguments. As in the two-particle example, $D(E)$ has only a right-hand cut, while $N(E)$ has only a left-hand cut. Thus we once more arrive at equations for N and D similar to (2.147, 148):

$$N(E) = \frac{1}{2\pi i} \int\limits_{L} \frac{dE'}{E' - E} \alpha(E')D(E') \ ; \tag{2.151}$$

$$D(E) = 1 - \frac{E}{\pi} \int\limits_{R} \frac{dE' K(E)' N(E')}{E'(E' - E)} \ , \tag{2.152}$$

where L and R are the integration contours along the left- and right-hand cuts, respectively. We shall find the step on the left-hand cut $\alpha(E)$ approximately, by using the diagram in Fig. 2.8 showing the singularity nearest to the physical region. Introducing the dimensionless variables and parameters $Z = \frac{E}{\varepsilon} = \frac{3k^2}{2\varepsilon m}$

(ε is the binding energy of the deuteron), $\gamma = (1 - r\sqrt{\varepsilon m})^{-1} \approx 1.65$ (r is the effective radius in the triplet state) we obtain the nonrelativistic limit

$$B(Z) = \frac{\tau 2\gamma}{Z\sqrt{3}} \ln \frac{1 + 3Z}{1 + Z/3} , \quad \tau = \begin{cases} +1 , & \text{spinless case;} \\ -1 , & \text{quartet nd-scattering} \end{cases} \qquad (2.153)$$

Thus, the left-hand cut is situated within the interval $-1/3 > Z > -3$. Equations (2.151, 152) in this notation assume the final form

$$D(Z) = 1 - \frac{Z}{\pi} \int_0^\infty dx \frac{\sqrt{x}N(x)}{x(x - Z)} ; \qquad (2.154)$$

$$N(Z) = \frac{1}{\pi} \int_{-3}^{-1/3} dx \frac{D(x)\mathrm{Im}\,B(x)}{x - Z} . \qquad (2.155)$$

Substituting (2.155) into (2.154) and integrating over x from 0 to infinity we obtain

$$D(Z) = 1 - i\sqrt{Z}N(Z) + \frac{iZ}{\pi} \int_{-3}^{-1/3} dy \frac{D(y)\mathrm{Im}\,B(y)}{\sqrt{y}(y - Z)} . \qquad (2.156)$$

Equation (2.156) can be solved approximately by the simple change of variables [45]

$$1/\sqrt{-Z} \approx C/(Z - a) + C_1 . \qquad (2.157)$$

In the integration region in (2.156) the approximation (2.157) leads to an uncertainty of about 2%. The constants C, C_1, and a are found from the requirement that the left and right-hand sides of (2.157) intersect at the points $Z = -3, -1, -1/3$. As a result, we obtain

$$a = \frac{3 - \sqrt{3}}{3\sqrt{3} - 1} \approx 0.3021 ; \quad C = -\frac{16}{(3\sqrt{3} - 1)^2} \approx -0.9095 ; \quad C_1 = a .$$

In the approximation (2.157) the equation (2.156) is solved with the function $N(Z)$ having only a left-hand cut; thus we find

$$D(Z) = 1 - i\sqrt{Z}N(Z) + \frac{CZ}{Z - a}[N(Z) - N(a)] + C_1 Z N(Z) . \qquad (2.158)$$

Clearly, on the left-hand cut (2.158) becomes

$$D(Z) \approx 1 - \frac{Z}{Z - a}CN(a) . \qquad (2.159)$$

Substituting (2.159) into (2.155) we obtain

$$N(Z) = B(Z) - \frac{CN(a)}{Z-a}[ZB(Z) - aB(a)] \ . \tag{2.160}$$

Allowing Z to approach a in (2.160) we find the unknown constant $N(a)$:

$$N(a) = B(a)/\{1 + C[B(a) + aB'(a)]\} \ . \tag{2.161}$$

Since $D(0) = 1$, for the quartet nd-scattering length ^4a we get

$$^4a = -\sqrt{(3/4)\varepsilon m}N(0) \approx 6.3 \,\text{fm} \ . \tag{2.162}$$

The experimental value of ^4a is 6.35 fm. Thus, the basic idea behind the approximation $\alpha(E) \approx B(E)$ is justified. Unfortunately, in the case of the doublet nd-scattering this simple choice for $\alpha(E)$ is insufficient.

Fig. 2.9. "Direct" diagram for an intermediate three-particle state, showing the main
contribution to the unitarity relation

Let us now see what the consequences for the spectrum of a three-particle system when starting with the three-particle unitarity relation for the amplitude $3 \rightarrow 3W(\boldsymbol{k}, \boldsymbol{p}, \boldsymbol{k}', \boldsymbol{p}', Z)$ are [28]. We introduce the partial amplitude $C_L(p, p', Z)$:

$$W(\boldsymbol{k}, \boldsymbol{p}, \boldsymbol{k}', \boldsymbol{p}') = \sum_L (2L+1) P_L(\cos \theta_{pp'}) t(k) C_L(p, p', Z) t(k') \ ,$$

where $t(k)$ is the S-harmonic of the pair T-matrix on the mass shell.

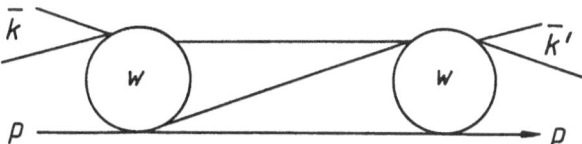

Fig. 2.10. "Exchange" diagram for an intermediate three-particle state, showing the main contribution
to the unitarity relation

The main contribution to the unitarity relation for the amplitude C_L is shown in two diagrams (Figs. 2.9, 10) in which, obviously, the amplitudes C_L are within the integral. It can be shown [46] that a good approximation for $C(p, p', Z)$ is the function

$$C\left(p, \sqrt{\frac{3}{4}mZ}, Z\right) = C(Z)\chi(p, Z), \chi\left(\sqrt{\frac{4}{3}mZ}, Z\right) = 1 \ , \quad (L=0), (2.163)$$

i.e., $C(Z) \equiv C\left(\sqrt{\frac{4}{3}mZ}, \sqrt{\frac{4}{3}mZ}, Z\right)$ is the scattering amplitude of a particle on a pair of particles with zero relative momentum (since $\frac{3}{4}p^2 + k^2 = mZ$). As a result, for the step in $C(Z)$ we obtain

$$\Delta C(Z) = -|C(Z)|^2[\varrho_d(Z) + \varrho_c(Z)]\theta(Z) , \tag{2.164}$$

where

$$\varrho_d(Z) = \frac{64\pi^3}{3\sqrt{3}}m \int\limits_0^{\sqrt{mZ}} q^2 dq \sqrt{mZ - q^2} \left|\chi\left(\sqrt{\frac{4}{3}(mZ - q^2)}; Z\right)\right|^2 |t(q)|^2 ; \tag{2.165}$$

$$\varrho_c(Z) = \frac{64\pi^3}{3\sqrt{3}}m \int dq_1^2 dq_2^2 \theta(1 - |u|)\chi\left(\sqrt{\frac{3}{4}(mZ - q^2)}; Z\right) t(q_1)t^*(q_2)\chi^*$$
$$\times \left(\sqrt{\frac{4}{3}(mZ - q^2)}; Z\right) \tag{2.166}$$

where $u = (mZ - p_1^2 - p_2^2)/p_1 p_2$; $p_i = \sqrt{\frac{4}{3}mZ - q_i^2}$ $(i = 1, 2)$. For $t(k)$ we shall use the approximate effective radius, i.e.,

$$t(k) = -\frac{1}{2\pi^2 m} \frac{1}{-a^{-1} + (1/2)r_0 k^2 - ik} . \tag{2.167}$$

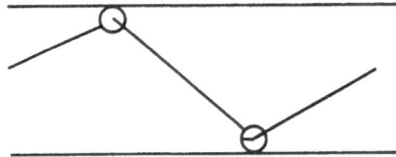

Fig. 2.11. The first-order Born amplitude $B^{(1)}$

In this case one must, as a jump on the dynamic cut, take the step in the second-order Born amplitude:

$$B^{(2)}(p, p', Z) = \int d^3 q B^{(1)}(p, q, Z) t\left(Z - \frac{3}{4}\frac{q^2}{m}\right) B^{(1)}(q, p', Z) , \tag{2.168}$$

where $B^{(1)}$ is the first-order Born amplitude shown in Fig. 2.11. This term itself in the approximation (2.163) yields no contribution to the unitary relation. The amplitude $B^{(2)}(p, p', Z)$ exhibits a logarithmic branch cut in Z from zero to negative infinity, and its jump is

$$\Delta B^{(2)}(Z) = -\frac{16\sqrt{3}\pi}{Z} \begin{cases} \frac{\pi}{12} - \sqrt{\frac{\varepsilon}{\varepsilon+Z}}\frac{\pi}{12}\left(\frac{\sqrt{\varepsilon}-\sqrt{-Z}}{\sqrt{\varepsilon}+\sqrt{-Z}}\right)^{1/2}, & Z > -\varepsilon; \\ \frac{\pi}{12} - \frac{1}{2}\sqrt{\frac{\varepsilon}{-\varepsilon-Z}}\ln\frac{\cot\frac{\pi}{12}+\left(\frac{\sqrt{-Z}-\sqrt{\varepsilon}}{\sqrt{-Z}+\sqrt{\varepsilon}}\right)^{1/2}}{\cot\frac{\pi}{12}-\left(\frac{\sqrt{-Z}-\sqrt{\varepsilon}}{\sqrt{-Z}+\sqrt{\varepsilon}}\right)^{1/2}}, & Z < -\varepsilon, \end{cases}$$

$$(2.169)$$

where $\varepsilon = 1/a^2 m$; a is the pair scattering length.

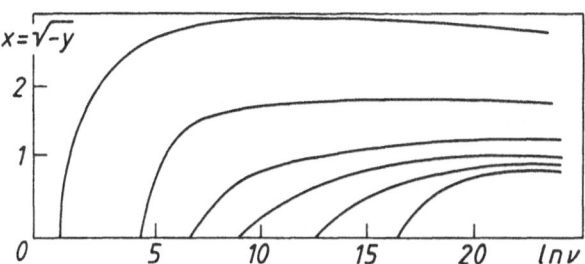

Fig. 2.12. Dependence of the trajectories of energy levels on the parameter $\nu = 2\frac{|a|}{r_0}$

Introducing the dimensionless quantities $Y = (1/4)mr_0^2 Z$, $\nu = 2|a|/r_0$ and taking into account the behavior of $C(Z)$, we obtain for $D(Z)$, as usual in the N/D-method, an analogue for (2.156):

$$D(y) = 1 + \frac{1}{\pi^2}\int_{-\infty}^{0} \Omega(y',\nu)l(y,y')D(y')dy', \qquad (2.170)$$

where

$$l(y,y') = (y,-y_0)\int_{0}^{\infty} \frac{r_d(y'') + r_c(y'')}{(y''-y')(y''-y)(y''-y_0)}dy'',$$

while Ω is determined by the step $\Delta B^{(2)}$. It is easy to show that for a finite ν (2.170) is a Fredholm equation and, thus, has a unique solution. In Fig. 2.12 trajectories $x = \sqrt{-y}$ of zeros of $D(y)$, i.e., trajectories of the energy levels as functions of $\ln\nu$ are presented.

The spectrum obtained possesses two remarkable properties:

1) If the pair scattering length tends to infinity, i.e., $|a| \rightarrow \infty$, then the parameter ν, by definition, also tends towards infinity. In this case, as can be seen from Fig. 2.12, the energy levels of the system accumulate in the vicinity of the point $Z = 0$, i.e., the Efimov effect occurs.

2) If the range of interaction r_0 tends to zero (for a finite $|a|$), the parameter ν again tends to infinity. As can be seen from Fig. 2.12, the trajectories of all

energy levels reach the asymptotic values y_i. However, if one recalls that the dimensional energy value equals $Z = 4e/(mr_0^2)$, then one immediately obtains that, as r_0 approaches zero, each energy level tends towards minus infinity as $-1/r_0^2$, i.e., a collapse, also known as the Thomas phenomenon, occurs.

From a mathematical point of view, the increase of the number of three-particle bound states [or of zeros of the function $D(y)$] with increasing ν is due to the unlimited increase of the norm of the kernel of (2.170). Indeed, for small y and y' and for ν tending to infinity we obtain for the kernel the estimate

$$K(y, y') \sim \frac{1}{y'} \frac{y' \ln y' - y \ln y}{y' - y} \qquad (2.171)$$

and for the norm of the kernel $N = \int |K(y, y')|^2 dy\, dy'$ we have $N \sim \ln^2 \nu$ as ν approaches infinity. Hence it follows that in order for $D(y)$ on the left-hand of (2.170) to remain finite as ν approaches infinity, the function D occurring in the integrand must strongly oscillate to compensate for the increase of the kernel. Thus, we obtain the familiar relationship between the number of Efimov levels [zeros of the function $D(y)$] and ν:

$$n \sim \ln \nu = \ln 2 \frac{|a|}{r_0} \qquad (2.172)$$

Finally, we note that by taking into account only two-particle unitarity, i.e., only two-particle intermediate states, we are lead instead of to (2.170), to a Fredholm equation for the function D, the properties of the equation remaining unaltered as ν becomes infinite. Thus, for the existence of the two effects mentioned above it is necessary that the three-particle amplitude satisfy the condition of three-particle unitarity.

We also note that this approach has made possible a comparatively simple description of the scattering of a particle on a resonance [47].

2.10 Unitarization Based on the Evolutionary Coupling Constant (ECC) Method

The third method of unitarization of the approximate expression for the three-particle amplitude is ideologically extremely close to the unitarization of the approximate solutions of the Faddeev equations (see above).

The difference is that this third method is based not on dynamic equations in configuration space, but on the so-called ECC equations [48] (equations with an evolutionary coupling constant).

First, we shall briefly present the nonrelativistic version of this approach and then give several examples of its application to real physical systems.

We define the Hamiltonian of a few-body system:

$$H = H_0 + gV ; \qquad (2.173)$$

$$H|\psi_p^{(+)}\rangle = E|\psi_p^{(+)}\rangle \ . \tag{2.174}$$

For the time being we shall assume that in (2.173) H_0 stands for the free Hamiltonian (below we shall deal in greater detail with other possibilities of choosing H_0); V is the sum of all the interactions between particles; g is a scalar quantity varying within the interval $[0, 1]$. We shall be interested in the evolution of state vectors, i.e., the solutions of (2.174), as well as in the evolution of matrix elements of various operators involving these solutions, depending on the variation of the parameter g. To this end we differentiate (2.174) with respect to g:

$$(H - E)\frac{d}{dg}|\psi_p^{(+)}\rangle = \left(\frac{dE}{dg} - V\right)|\psi_p^{(+)}\rangle \ . \tag{2.175}$$

For simplicity, we shall at first restrict ourselves to the scattering problem in which the energy of the system E is an external parameter, therefore, $dE/dg = 0$. Then, from (2.175) we obtain for the state vector a differential equation with respect to the coupling constant g:

$$\frac{d}{dg}|\psi_p^{(+)}\rangle = -G(E)V|\psi_p^{(+)}\rangle \tag{2.176}$$

or

$$\frac{d}{dg}|\psi_p^{(+)}\rangle = \int \frac{d\mathbf{p}'}{(2\pi)^3} \frac{V_{p,p'}|\psi_{p'}^{(+)}\rangle}{E_{p'} - E + i\varepsilon} \ , \tag{2.177}$$

where $V_{p'p} \equiv \langle\psi_{p'}^{(+)}|V|\psi_p^{(+)}\rangle$ is the matrix element of the interaction operator over the solutions of (2.174). The boundary condition relative to g for our choice of H_0 is obvious: when $g = 0$, we must choose for $|\psi_p\rangle$ a free solution. Using (2.177) it is possible to obtain an equation for the matrix element of any operator O by differentiation:

$$\frac{d}{dg}O_{p'p} = \int \frac{d\mathbf{q}}{(2\pi)^3}\left[\frac{V_{p'q}O_{qp}}{E_{p'} - E_q - i\varepsilon} + \frac{O_{p'q}V_{qp}}{E_p - E_q + i\varepsilon}\right] \tag{2.178}$$

Thus, for instance, if $O = V$, we obtain the nonlinear equation

$$\frac{dV_{p'p}}{dg} = \int \frac{d\mathbf{q}}{(2\pi)^3} V_{p'q}V_{qp}[E_{p'} - E_q - i\varepsilon)^{-1} + (E_p - E_q + i\varepsilon)^{-1}] \ . \tag{2.179}$$

Let us derive the equation for the S-matrix. We use the definition

$$S_{pp'} = \langle\psi_p^{(-)}|\psi_{p'}^{(+)}\rangle \tag{2.180}$$

or, which is the same,

$$\langle\psi_p^{(-)}| = \int \frac{d\mathbf{q}}{(2\pi)^3} S_{pq}\langle\psi_q^{(+)}| \ . \tag{2.181}$$

Using (2.177, 181) we obtain the relation

$$\frac{d}{dg}\langle\psi_p^{(-)}| = \int \frac{dq_1 dq_2}{(2\pi)^6} \frac{S_{pq_1} V_{q_1 q_2} \langle\psi_{q_2}^{(+)}|}{E_p - E_{q_2} + i\varepsilon} .$$

(2.182)

Combining (2.177, 182), and the definition of the S-matrix (2.180) we ultimately get

$$\frac{dS_{pp'}}{dg} = -2\pi i \int \frac{dq}{(2\pi)^3} S_{pq} V_{qp'} \delta(E_q - E_{p'}) .$$

(2.183)

Thus, to find the scattering matrix $S_{pp'}$, it is necessary to solve a set of two equations (2.179, 183). Separating out in (2.183) the angular variables we obtain the ECC equation for the scattering phase:

$$\frac{d\delta_l}{dg} = \frac{k\mu}{2\pi} V_k^l ,$$

(2.184)

where V_k^l is the partial harmonic of the matrix element $V_{kk'}$ for $k = k'$.

Thus, for any approximate solution for $V_{pp'}$ the S-matrix remains unitary, since, owing to (2.184), the approximation is in the phase.

We shall now demonstrate the application of the ECC method to the three-body problem. We shall first deal with nd-scattering in the quartet state at low energies (in this case at an energy lower than the deuteron breakup threshold).

Since three-particle scattering may give rise to disconnected diagrams giving no contribution to the physical process, we shall single out their contribution to the matrix element V determining the scattering amplitude:

$$\langle pq|V|p'q'\rangle = \langle pq|V^C|p'q'\rangle + \langle pq|U|p'q'\rangle ,$$

(2.185)

where $|pq\rangle \equiv |\psi_{pq}^+\rangle$ is the scattering state. Here U corresponds to the disconnected part, i.e., $U = \sum v_i$; v_i is the pair potential. Ignoring the disconnected terms we obtain for V^C

$$\begin{aligned}
\frac{d}{dg}\langle pq V^c p'q'\rangle = \int \frac{dp'' dq''}{(2\pi)^6} \langle pq|[(U|p''q''\rangle\langle p''q''|U)^c \\
+ U|p''q''\rangle\langle p''q''|V^c + V^c|p''q''\rangle\langle p''q''|U \\
+ V^c|p''q''\rangle\langle p''q''|V^c]|p'q'\rangle[(E_{pq} - E_{p'q'} - i\varepsilon)^{-1} \\
+ (E_{p'q'} - E_{p''q''} + i\varepsilon)^{-1}] ,
\end{aligned}$$

(2.186)

where $(UU)^c = \sum_{i \neq j} v_i v_j$; E_{pq} is the energy of the three-particle state.

The form of the right-hand side suggests the method of solving (2.186) – iteration in the right-hand side in powers of V^c. In the case of nd-scattering the exchange diagram in Fig. 2.13 corresponds to the zeroth iteration, i.e., (2.186) becomes

$$\frac{d}{dg}\langle nd|V|nd\rangle = 2\frac{|\langle d|V|np\rangle|^2}{q^2 + \kappa^2} , \quad \kappa^2 = -|\varepsilon_d|2\mu .$$

(2.187)

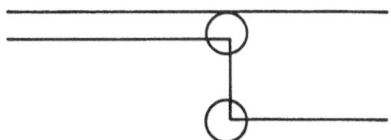

Fig. 2.13. Modified exchange diagram for elastic nd-scattering

The explicit expression for q is given below. It is easy to verify that for the separable pair NN-potential V the following relation holds:

$$|\langle d|V|np\rangle|^2 = \langle d|V|d\rangle \langle np|V|np\rangle = \frac{4\pi}{q}\frac{d\delta_l^{NN}(q\kappa)}{d\kappa^2} , \qquad (2.188)$$

where δ^{NN} is the phase of the triplet NN-scattering.

In deriving (2.188) we made use of the Gell–Mann–Low relation readily obtained from (2.175)

$$\langle d|V|d\rangle = d\varepsilon/dg , \qquad (2.189)$$

and substituted differentiation with respect to κ^2 for differentiation with respect to g, and expressed the two-particle matrix element $\langle np|V|np\rangle$ through the scattering phase

$$\langle k|Vk'\rangle_l = -4\pi\left(\frac{1}{kk'}\frac{d\delta_l^{NN}(k)}{dg}\frac{d\delta_l^{NN}(k')}{dg}\right)^{1/2}$$
$$\times \exp\left\{i[\delta_l^{NN}(k) - \delta_l^{NN}(k')]\right\} . \qquad (2.190)$$

Further, by using an equation similar to (2.184) for the nd-scattering phase,

$$d\delta_l^{nd}/dg = -(k/3\pi)\langle nd|V|nd\rangle \qquad (2.191)$$

and (2.187, 188), we finally obtain for the nd-scattering phase

$$\delta_l^{nd} = -\left[\frac{8k}{3q}(q^2 + \kappa^2)\int_{\kappa^2}^{\infty}\frac{d\kappa'\delta^{NN}(q, \kappa')}{(q^2 + \kappa'^2)^2}\right]_l , \qquad (2.192)$$

where $q = k(5/4 + \cos\theta)^{1/2}$ and the index l in the right-hand side of (2.192) corresponds to the l-th partial wave of the expression in square brackets.

Expression (2.192) leads to the expansion in the small quantity $\kappa/\beta \approx 0.16$ for the quartet nd-scattering length

$$a_{3/2} = \frac{16}{9\kappa}\left[1 - \frac{3}{4}\frac{\kappa}{\beta} + \frac{3}{4}\frac{\kappa^2}{\beta} + \ldots\right] \approx 6.75 \text{ fm} . \qquad (2.193)$$

This quartet scattering length agrees quite well with the experimental value of $a_{3/2} = 6.35$. The quartet effective radius is also in good agreement with (2.192) [49].

As in the previous method of unitarization of the three-particle amplitude, in the present approach it turns out to be impossible to describe the doublet nd-scattering length. The reason for this is the same: the insufficiency of taking into account only the exchange (Fig. 2.13).

Application of this unitarization method is more effective in the case of pion-nuclear interactions. We shall now discuss pion-nuclear scattering at low energies [50].

As is usual in this method, we consider the Hamiltonian of the pion-nucleus system:

$$H = H_0 + h + gV_\pi \; . \tag{2.194}$$

Here H_0 is the kinetic energy operator of the relative pion-nucleus motion; h is the Hamiltonian of the nucleus; $V_\pi = \sum_{i=1}^{N} V_{\pi N_i}$ is the interaction potential between the pion and the nucleus consisting of A nucleons. The physical situation clearly corresponds to the Hamiltonian (2.194) for $g = 1$.

We now introduce the state vectors $|\mu\rangle$ which are eigenvectors of the Hamiltonian (2.194) and the state vectors $|\mu\rangle$ that are eigenvectors of the Hamiltonian $\tilde{H} = H_0 + h$. From the equation for the scattering length (2.191) we readily obtain the equation for the pion-nucleus scattering length:

$$da_{\pi A}/dg = -(\mu_{\pi A}/2\pi)(\pi A|V|\pi A) \; . \tag{2.195}$$

To calculate the right-hand side of (2.195) as a function of g we shall not solve a nonlinear equation for the matrix elements as in the case of nd-scattering, but shall make use of a more simple procedure. Instead of the potential V_π, we introduce the transition operator through the relation

$$T|\pi A\rangle = gV_\pi|\pi A\rangle \; . \tag{2.196}$$

Then, for the desired matrix element we have

$$(\pi A|V_\pi|\pi A) = \frac{1}{g}(\pi A|T|\pi A) \; . \tag{2.197}$$

We now compute the right-hand side of the relation (2.197) in the approximation linear in the elementary πNT-matrix. In this approximation the pion and the nucleus in the final state are to be considered free, i.e., it is necessary to substitute $|\pi A\rangle$ for the state vector $|\pi A\rangle$, since by including the interaction one takes into account the higher orders in the expansion of the right-hand side of (2.197) in powers of $t_{\pi N}$.

Thus, in this approximation (2.196) takes the form

$$\frac{da^{(1)}}{dg} = -\frac{\mu_{\pi A}}{2\pi}\frac{1}{g}\langle\pi A|T^{(1)}|\pi A\rangle \; . \tag{2.198}$$

The matrix element in the right-hand side of (2.198) in the impulse approximation is

$$\langle \pi A | T^{(1)}(E_\pi \approx 0) | \pi A \rangle = \sum_{i=1}^{N} \int \frac{d\boldsymbol{k}_1}{(2\pi)^3} \cdots \frac{d\boldsymbol{k}_A}{(2\pi)^3} (2\pi)^3 \delta(\boldsymbol{k}_1 + \ldots + \boldsymbol{k}_A)$$

$$\times \psi_A(\boldsymbol{k}_1, \ldots \boldsymbol{k}_A) \langle \overset{\circ}{\boldsymbol{k}}_i | t_{\pi N_i} | \overset{\circ}{\boldsymbol{k}}_i \rangle$$

$$\times \psi_A \times (\boldsymbol{k}_1, \ldots \boldsymbol{k}_A) , \tag{2.199}$$

where $\langle \overset{\circ}{\boldsymbol{k}}_i | t_{\pi N_i} | \overset{\circ}{\boldsymbol{k}}_i \rangle$ is the matrix element of the t-matrix in the center-of-mass system of the pion and the i-th nucleon, and $\overset{\circ}{\boldsymbol{k}}_i = (\mu_\pi / m_N + \mu_\pi) \boldsymbol{k}_i$. In the static limit, i.e., when $\mu_\pi / m_N \ll 1$, and neglecting the contribution of the P-wave to the πN-interaction, one may replace the matrix element $\langle \overset{\circ}{\boldsymbol{k}}_i | t_{\pi N} | \overset{\circ}{\boldsymbol{k}}_i \rangle$ in the integral (2.199) by its value at $k_i = 0$. With allowance for the isotopic structure of the πN-interaction for $t_{\pi N_i}$ we have

$$t_{\pi N_i} = t_0 + t_1(\boldsymbol{T}\tau_i) , \tag{2.200}$$

where $\boldsymbol{T}$ is the isotopic spin of the pion; τ_i is the isotopic spin of the ith nucleon; t_0 and t_1 are the isoscalar and isovector components of the πN-amplitude, respectively. Ultimately, we obtain for the amplitude (2.199)

$$\langle \pi A | T^{(1)} | \pi A \rangle = A t_0(g) + 2 \langle \boldsymbol{T}\boldsymbol{I} \rangle t_1(g) = A t_0(g) \pm (Z - N) t_1(g) . \tag{2.201}$$

Here $\pm$ corresponds to scattering of $\pi^\pm$-mesons; $\boldsymbol{I}$ represents the isospin operator of the nucleus; Z and N are the numbers of protons and neutrons, respectively; $A = Z + N$. Using (2.201) the equation for the scattering length is

$$da_{\pi A}^{(1)}/dg = (1/g)[A a_0(g) \pm (Z - N) a_1(g)] , \tag{2.202}$$

where

$$a_0(g) = -\frac{\mu_{\pi N}}{2\pi} t_0 = \tfrac{1}{3}[a_{1/2}(g) + 2 a_{3/2}(g)] ; \tag{2.203}$$

$$a_1(g) = -\frac{\mu_{\pi N}}{2\pi} t_1 = -\tfrac{1}{3}[a_{1/2}(g) - a_{3/2}(g)] ; \tag{2.204}$$

$a_{1/2}(g)$ and $a_{3/2}(g)$ are the πN-scattering lengths as functions of the coupling constant g, corresponding to states with total isospins $1/2$ and $3/2$. It is easy to check that for the separable πN-potential

$$V_{\pi N}^0 = \lambda v(k) v(k') , \quad v(k) = 1/(k^2 + \gamma^2) . \tag{2.205}$$

By solving the two-particle ECC equation for the scattering length we obtain

$$a_i(g) = g a_i^{\text{exp}} / \{1 - (\gamma/2)(g - 1) a_i^{\text{exp}}\} , \quad i = \tfrac{1}{2}, \tfrac{1}{3} , \tag{2.206}$$

where a_i^{exp} are the corresponding experimental πN-scattering lengths.

Solving (2.202) for the pion-nucleus scattering length we obtain

$$a_{\pi A}^{(1)} = A \bar{a}_0 \pm (Z - N) \bar{a}_1 ; \tag{2.207}$$

$$\bar{a}_0 = \frac{2}{3\gamma}[\ln|1 + d_{1/2}| + 2\ln|1 + d_{3/2}|] , \qquad (2.208)$$

$$\bar{a}_1 = -\frac{2}{3\gamma}\ln\left|\frac{1 + d_{1/2}}{1 + d_{3/2}}\right| , \quad d_i = \frac{\gamma}{2}a_i . \qquad (2.209)$$

From (2.207–209) it follows that (2.207) already contains all the powers of the elementary πN-scattering length. The values found from (2.207) differ strongly from the non-unitarized values and at the same time are close to the experimental scattering lengths for a great variety of light nuclei from deuterium to sodium.

We shall now try to determine the matrix element $(\pi A|V_\pi|\pi A)$ which ultimately yields the π-nucleus scattering length more carefully. The correction to be made is that this matrix element will be determined by also including terms quadratic in $t_{\pi N}$, and not only the linear ones. For simplicity, we shall consider the case of πd-scattering. We represent the matrix element $(\pi d|T|\pi d)$ in the form

$$(\pi d, \mathbf{k}'|T|\pi d, \mathbf{k}) = \int \frac{d\mathbf{k}_1 d\mathbf{q}_1}{(2\pi)^6}(\pi d, \mathbf{k}'|\mathbf{k}_1 \mathbf{q}_1)\langle\mathbf{k}_1 \mathbf{q}_1|T|\pi d, \mathbf{k}) , \qquad (2.210)$$

where $(\pi d, \mathbf{k}'|\mathbf{k}_1, \mathbf{q}_1) \equiv \psi_{\mathbf{k}_1}^*(\mathbf{k}_1, \mathbf{q}_1)$ is the wave function of the πd-system, for which at low energies one can utilize the approximate expression

$$\psi_{\mathbf{k}'}^*(\mathbf{k}_1 \mathbf{q}_1) = (2\pi)^3 \delta(\mathbf{k}_1 - \mathbf{k}')\varphi_d(\mathbf{q}_1)$$
$$+ \left[Z' - \frac{k_2^2}{2\mu_{\pi d}} - \frac{q_1^2}{m_N}\right]^{-1}$$
$$\times \int \frac{d\mathbf{q}_1'}{(2\pi)^3}\varphi_d(\mathbf{q}_1')\langle\mathbf{k}', \mathbf{q}_1'|T^*|\mathbf{k}_1 \mathbf{q}_1) . \qquad (2.211)$$

By using (2.211), (2.210) can be rewritten

$$(\pi d, \mathbf{k}|T|\pi d, \mathbf{k}) = \langle\pi d, \mathbf{k}'|T|\pi d, \mathbf{k})$$
$$+ \int \frac{d\mathbf{q}' d\mathbf{q}}{(2\pi)^6}\varphi_d(\mathbf{q}')\langle\mathbf{k}, \mathbf{q}'|T^*G_0(E - i\varepsilon)T|\mathbf{k}\mathbf{q}\rangle\varphi_d(\mathbf{q}). \qquad (2.212)$$

In the approximation quadratic in $t_{\pi N}$ (2.212) is

$$(\pi d, 0|T^{(2)}|\pi d, 0) = -\int \frac{d\mathbf{k}_1 d\mathbf{q}_1}{(2\pi)^6}\left\{\varphi_d\left(\mathbf{q}_1 + \frac{\mathbf{k}_1}{2}\right)t_{\pi N_1}t_{\pi N_1}\varphi_d\left(\mathbf{q}_1 + \frac{\mathbf{k}_1}{2}\right)\right.$$
$$+ \varphi_d^*\left(\mathbf{q}_1 - \frac{\mathbf{k}_1}{2}\right)t_{\pi N_2}t_{\pi N_2}\varphi_d\left(\mathbf{q}_1 - \frac{\mathbf{k}_1}{2}\right)$$
$$+ 2\left[\varphi_d^*\left(\mathbf{q}_1^+ + \frac{\mathbf{k}_1}{2}\right)t_{\pi N_1}t_{\pi N_2}\varphi_d\left(\mathbf{q}_1 - \frac{\mathbf{k}_1}{2}\right)\right.$$
$$+\varphi_d^*\left(\mathbf{q}_1 - \frac{\mathbf{k}_1}{2}\right)t_{\pi N_2}t_{\pi N_1}\varphi_d\left(\mathbf{q}_1 + \frac{\mathbf{k}_1}{2}\right)\left.\right]\right\}$$
$$\times \left(\frac{\alpha^2}{m_N} + \frac{k_1^2}{2\mu_{\pi d}} + \frac{q_1^2}{m_N}\right)^{-1} . \qquad (2.213)$$

The first two terms in braces, which will be denoted by $T^{(2)}_{N_1 N_1}$ and $T^{(2)}_{N_2 N_2}$, in the static limit are equal to

$$T^{(2)}_{N_1 N_1} + T^{(2)}_{N_2 N_2} = -2(t_0^2 + 2t_1^2)\gamma \mu_{\pi N}/4\pi \ . \tag{2.214}$$

The contribution to the scattering length from (2.214) is determined by

$$d\tilde{a}^{(2)}_{\pi l}/dg = (\gamma/g)[a_0^2(g) + 2a_1^2(g)] \ . \tag{2.215}$$

Solving (2.215) we obtain

$$\tilde{a}^{(2)}_{\pi d} = (4/3\gamma)(d_{1/2} + 2d_{3/2} - \ln|1 + d_{1/2}| - 2\ln|1 + d_{3/2}|) \ . \tag{2.216}$$

The remaining term in (2.213) assumes in the static limit the form

$$T^{(2)}_{N_1 N_2} = -2\frac{\mu_{\pi N}}{2\pi} \langle \frac{1}{r} \rangle (t_0^2 - 2t_1^2) \ , \tag{2.217}$$

where the mean inverse radius is determined by

$$\langle \frac{1}{r} \rangle = \int \frac{dk dq}{(2\pi)^6} \varphi_d \left(q + \frac{k}{2}\right) \frac{1}{k^2} \varphi_d \left(q - \frac{k}{2}\right) \ . \tag{2.218}$$

The contribution of $T^{(2)}_{N_1 N_2}$ to the πd-scattering length is given by

$$\frac{da^{(2)}_{\pi d}}{dg} = \frac{4}{g} \langle \frac{1}{r} \rangle [a_0^2(g) - 2a_1^2(g)] \tag{2.219}$$

and equals

$$a^{(2)}_{\pi d} = \frac{16}{9\gamma^2} \langle \frac{1}{r} \rangle \Big\{ -(d_{1/3} - \ln|1 + d_{1/2}|) + 2(d_{3/2} - \ln|1 + d_{3/2}|)$$
$$+ \frac{8}{d_{1/2} - d_{3/2}} [d_{3/2}(1 + d_{1/2})\ln|1 + d_{1/2}| - d_{1/2}$$
$$\times (1 + d_{3/2})\ln(|1 + d_{3/2}|] \Big\} \ . \tag{2.220}$$

Thus, for the πd-scattering length

$$a_{\pi d} = a^{(1)}_{\pi d} + a^{(2)}_{\pi d} + \tilde{a}^{(2)}_{\pi d} \ . \tag{2.221}$$

In Table 2.2 the scattering length (2.221) is compared with the length found by solving the Faddeev equations for various sets of parameters of πN-potentials (a, b, c, d).

Table 2.2.

Scattering length	a	b	c	d
a_F	−0.074	−0.030	−0.061	−0.045
a(2.221)	−0.072	−0.029	−0.059	−0.044

As we see from the Table, (2.221) yields practically the exact value of the πd-scattering length.

For completeness, we present the expression for the πd-scattering length including the terms quadratic in $t_{\pi N}$:

$$a_{\pi A}^{(2)} = \frac{8}{9\gamma^2}\langle\frac{1}{r}\rangle\{(P_1 + P_2 + P_3)(d_{1/2} - \ln|1 + d_{1/2}|)$$
$$+ (4P_1 - 2P_2 + P_3)(d_{3/2} - \ln|1 + d_{3/2}|)$$
$$+ (4P_1 + P_2 - 2P_3)(d_{1/2} - d_{3/2})^{-1}$$
$$\times [d_{3/2}(1 + d_{1/2})\ln|1 + d_{1/2}| - d_{1/2}(1 + d_{3/2})\ln|1 + d_{3/2}|]\} , \quad (2.222)$$

where

$$P_1 = A(A + 1); \quad P_2 = 2(A - 1)(Z - N); \quad P_3 = (Z - N)^2 - 2A ;$$

$$\tilde{a}_{\pi A}^{(2)} = \frac{2A}{3\gamma}(d_{1/2} + 2d_{3/2} - \ln|1 + d_{1/2}| - 2\ln 1 + d_{3/2}|)$$

$$- (Z - N)\frac{2}{3\gamma}\left(-d_{1/2} + d_{3/2} + \ln|\frac{1 + d_{1/2}}{1 + d_{3/2}}|\right) .$$

Let us now raise the question: what physical information can be obtained from (2.207, 222) for the π-nucleus scattering lengths? In particular, what can be found out about the properties of a nuclear system by studying the π-nucleus scattering length? From the presented formulas one can see that the structure of the nucleus manifests itself in two places. First, already in the approximation linear in $t_{\pi N}$, the matrix element of the transition operator $\langle\pi A|T|\pi A\rangle$ begins to depend on the nuclear wave functions if one drops the static ($\mu_\pi/m_N \ll 1$) approximation. However, the contribution of the summands which are proportional to μ_π/m_N lies within the uncertainty with which the contribution of the static term can be calculated. This is due to the uncertainty in the value of the isoscalar πN-scattering length $a_0^{\pi N}$. Thus, in order to apply the formulas of the linear approximation to the structure of the nucleus a significant enhancement of the reliability of the value of $a_0^{\pi N}$ is necessary, i.e., it is necessary for the uncertainty to be less than μ_π/m_N, which means that is has to be better than 13 %.

Second, the structure of the nucleus manifests itself when the terms quadratic in $t_{\pi N}$ are taken into account in the form of the nuclear matrix element such as $\langle\exp(\alpha^*r)/r\rangle$ and $\langle1/r\rangle$ where $\alpha^* = \alpha(2\mu_{\pi A}/m_N)^{1/2}$; $\alpha = \sqrt{2\mu\varepsilon}$; ε is the binding energy of the nucleus. However, calculations of the πA-scattering length for different sets of πN data lead to significantly higher discrepancies than the contribution of the terms quadratic in $t_{\pi N}$. Thus, in order to extract information on the structure of the nucleus from the data on the πA-scattering, it is necessary to select the most appropriate elementary πN data.

We shall now discuss the possible generalizations of the ECC method [51]. As has already been noted, the starting point of the idea behind the ECC method is the assertion that the description of the evolution of a dynamic system in time is equivalent to a description of the evolution in the coupling constant. Hence it is clear that the "natural" boundary condition necessary to solve the differential equations of the ECC method must be the condition at the point $g = 0$, where g is the coupling constant. Thus, in this case the evolution of the system starts with the free motion of all the particles belonging to the system.

The essence of the generalization is that we assume that the evolution of the system does not start from free motion ($g = 0$), but from a given value $g = g_0$ for which the exact solution of the problem is known, having been obtained by some other method. We shall illustrate these arguments by two examples of the three-body problem in the domain of the πd-scattering at zero pion energy.

Instead of the Hamiltonian of the $\pi 2N$ system, as usual in the ECC method, we now introduce the auxiliary Hamiltonian

$$H(g) = H_0 + V_\pi + gh .\tag{2.223}$$

Here H_0 is the kinetic energy of the relative motion of the pion and the center-of-mass of the two nucleons; $V_\pi = V_{\pi N_1} + V_{\pi N_2}$ is the interaction potential between the pion and the nucleons; h is the Hamiltonian of the $2N$ system; the coupling constant g is a parameter that varies within the interval $[0, 1]$.

Clearly, $H(1)$ represents the Hamiltonian of the real $\pi 2N$ system, while $H(0)$ is the solvable Hamiltonian for the case of separable πN-potentials. Technically, the solution of the Schrödinger equation with the Hamiltonian $H(0)$ is obtained in the same way as in the problem of an incident particle interacting with fixed centers. This is already so, since from (2.223) it is seen that in the Hamiltonian $H(0)$ no kinetic energy of the nucleons is present. We note that taking into account the term gh even approximately, represents simultaneously the contributions of the discrete and continuous spectra of the Hamiltonian of the target. This is the important feature which distinguishes this proposed method from the well-known approximations of the multiple-scattering theory, such as the optical model and others, in which it is usually difficult to take into account the contribution of the continuous spectrum of the Hamiltonian of the target.

Applying equations of the ECC method for the Hamiltonian (2.223) we shall determine the corrections to the πd-scattering length a_0 calculated exactly with the Hamiltonian $H(0)$ which, for simplicity, will henceforth be called the Hamiltonian of the problem with fixed centers.

The equation for the scattering length is

$$\frac{da}{dg} = -\frac{\mu}{2\pi} \langle \psi_{k'=0} | \hbar | \psi_{k=0} \rangle^c a(g = 0) = a_0 \tag{2.224}$$

where $|\psi_k\rangle$ is the eigenfunction of the Hamiltonian (2.223):

$$\langle \psi_{k'} | \hbar | \psi_k \rangle = \langle \psi_{k'} | h | \psi_k \rangle - E\delta(k - k') ;$$

the index c indicates the connected part of the matrix element; E is the total energy of the system, which in this is equal to the binding energy of the deuteron.

The matrix element on the right-hand side of (2.224) can be written

$$\langle \psi_{k'} | \hbar | \psi_k \rangle^c = h^{(1)} + h^{(2)} .$$

Here

$$h^{(1)} = \int \frac{dq_1 dq_2}{(2\pi)^6} \left[\varphi_d(q_1) \frac{\langle k q_1 | h | k' q_2 \rangle}{Z - k'^2/(2\mu) - q_2^2/(2m')} \right.$$
$$\times \eta(k', q_2; k) + \eta^*(k, q; k') \langle k q_1 | h | k' q_2 \rangle$$
$$\left. \times (Z'^* - k^2/2\mu - q_1^2/2m')^{-1} \varphi_d q_2) \right] ; \tag{2.225}$$

$$h^{(2)} = \int \frac{dk_1 dq_1 dk_2 dq_2}{(2\pi)^{12}} \eta^*(k_1 q_1; k')$$
$$\times \frac{\langle k_1 q_1 | h | k_2 q_2 \rangle}{(Z'^* - k_1^2/(2\mu) - q_1^2/(2m'))} \frac{\eta(k_2, q_2; k)}{(Z - k_2^2/(2\mu) - q_2^2/(2m))} , \tag{2.226}$$

where

$$m' = \frac{m}{g} ; \quad Z' = k^2/2\mu + \alpha^2/m' ;$$
$$\eta(k_1, q_1; k') = \int \frac{dq}{(2\pi)^3} \langle k_1 q_1 | T | k', q \rangle \varphi_d(q) ; \tag{2.227}$$

T is the transition operator for the $\pi 2N$ system; φ_d is the deuteron wave function.

In obtaining this representation of the matrix element we have again made use of the approximate representation for the wave function of the πd system:

$$\psi_{k'}(k_1, q_1) = (2\pi)^3 \delta(k' - k_1) \varphi_d(q_1)$$
$$+ \left(Z' - \frac{k_1^2}{2\mu} - \frac{q_1^2}{m'} \right)^{-1}$$
$$\times \int \frac{dq_1'}{(2\pi)^3} \langle k_1 q_1 | T | (Z') | k q_1' \rangle \varphi_d(q_1') . \tag{2.227'}$$

From the definition of h it follows that $h^{(1)}$ turns, in our case ($E = \varepsilon_d$), into zero. The expression of $h^{(2)}$ is found approximately by calculating the function $\eta(k_1, q_1; k)$ for the limit $g = 0$. (Following [52] one can verify that the function η can be expanded in a series in powers of g.) As a result, we obtain for $h^{(2)}$

$$h^{(2)} = a_0^2 (Y/2) \sqrt{g} , \tag{2.228}$$

where a_0 is the πd-scattering length at $g = 0$; $Y \simeq 0.05$. Substituting (2.228) into (2.224) we find, with the boundary condition taken into account,

$$a(g) = a_0 - \sqrt{g} a_0^2 Y . \tag{2.229}$$

For the value $g = 1$,

$$a_{\pi d} = -0.06 \, \text{fm} \tag{2.230}$$

Thus, as can be seen from (2.229), the correction to the πd-scattering length, found in the approximation of fixed centers, is quadratic in a_0 and its contribution is quite small, because a_0 is itself small for the πd system. The physical reason for such a situation is that the πN-interaction is small on a nuclear scale.

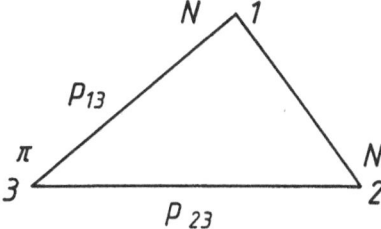

Fig. 2.14. Relative momenta in the $\pi 2N$ system: p_{13} is the relative momentum of the pion and the nucleon 1; p_{23} is the same for the nucleon 2

We shall now deal with the second example of generalization of the ECC method. The idea to be exploited here is the same as in the first example – the idea that the Schrödinger equation is solvable with the Hamiltonian $H(g)$ for a certain special value of $g = g_0$. Here $H(g)$ will be understood to be the Hamiltonian

$$H(g) = h_{13} + h_{23} + g h_{12} , \qquad (2.231)$$

where $h_{ij} = p_{ij}^2/2\mu_{ij} + V_{ij}$, $ij = (13), (23)$; $h_{12} = (p_{13}p_{23}/\mu_\pi) + V_{12}$. The adopted notation is clear from Fig. 2.14.

As in the previous case, $H(1)$ corresponds to the Hamiltonian of the physical $(\pi 2N)$ system. It can be shown that for a certain value of the constant $g = g_0$ the πd-scattering length will be equal to the sum of the elementary scattering lengths [53]. This value g is now the boundary value. To clarify the physical meaning of g_0 we rewrite the Hamiltonian (2.231) in the Jacobi variables $k = p_{13} + p_{23}$ and $q = (1/2)(p_{13} - p_{23})$:

$$H(k, q) = \frac{2m' + \mu'}{2m'\mu'} k^2 + \frac{q^2}{m^2} + V_{13} + V_{23} + g V_{12} , \qquad (2.232)$$

where

$$\mu' = \mu \hat{\pi}/g ; \quad m' = \mu_\pi/(1 - g + \mu_\pi/m) .$$

Thus, the Hamiltonian of the nucleon subsystem becomes

$$h_{NN} = q^2/m'(g) + g V_{12} . \qquad (2.233)$$

It turns out that at $g = g_0$ the Hamiltonian (2.223) has an eigenvalue equal to zero. From this condition we can determine the numerical value of g_0. Thus, the three-particle $H(g_0)$ describes interaction of an incident particle of mass μ_π/g_0 with the bound complex of two particles, the mass of each one of which equals $m'(g_0)$, while the binding energy of the complex is equal to zero.

Into the right-hand side of (2.224) one must substitute the matrix element of the operator

$$h_{12} = V_{12} + (1/\mu_\pi)(k^2/4 - q^2) , \qquad (2.234)$$

i.e., the quantity

$$\langle \psi_{k'} | h_{12} | \psi_k \rangle = \int \frac{dk_1 dq_1 dk_2 dq_2}{(2\pi)^{12}}$$
$$\times \psi_{k'}^*(k_1 q_1)\langle k_1 q_1 | h_{12} | k_2 q_2 \rangle \psi_k(k, q_2) . \tag{2.235}$$

In order to calculate (2.235) and solve (2.224) we once again make use of the representation of the wave function (2.227) and of the boundary condition

$$a(g_0) = (\mu_{\pi d}(g_0)/\mu_{\pi N})2a^0 , \tag{2.236}$$

where a_0 represents an isoscalar combination of the pion-nucleon scattering lengths, and $\mu_{\pi d}(g) = 2\mu_\pi/(1 + g + \mu_\pi/m_N)$.

We shall search for a solution of (2.224) in the form

$$a(g) \approx a(g_0) + (g - g_0)a^{(1)}(g_0) . \tag{2.237}$$

The approximation (2.237) is justified since for the NN-potential V_{12} of the separable triplet the numerical value of g_0 equals 0.744, i.e., it is close to unity. Thus, in (2.237) it is only necessary to find the constant $a^{(1)}(g_0)$, i.e., only the right-hand side of (2.224) at $g = g_0$. Ultimately, for $a(g)$ we obtain

$$a(g) = a(g_0)\left[1 + (g - g_0)\left(\frac{2}{g_0} + \frac{3}{2}\frac{1}{1 - g_0 + \mu_\pi/m_N}\right)\right] . \tag{2.238}$$

From (2.238) there follows:

$$a_{\pi d} = a(1) \simeq 2.64a(g_0) = -0.066 \, \text{fm} . \tag{2.239}$$

To demonstrate the quality of the estimates given by (2.230, 239) we compare them with the solution of the Faddeev equations and with the corresponding experimental value:

$$a_{\pi d}^F = -0.061 \, \text{fm} ; \quad a_{\pi d}^{exp} = (-0.073 \pm 0.030) \, \text{fm} ; \tag{2.240}$$

$$a_{\pi d} = -0.066 \, \text{fm} \quad \text{via (2.239)} ;$$

$$a_{\pi d} = -0.060 \, \text{fm} \quad \text{via (2.230)} .$$

Within the framework of the ECC method it is easy to estimate the contribution to the πd-scattering length from the so-called indirect interaction of particles 1 and 2, i.e., the contribution of the term $p_{12}p_{13}/\mu_\pi$. Calculation of $a(g)$ without this term gives

$$a(g) = a(g_0)g^2/g_0^2 . \tag{2.241}$$

Comparison with the estimate when the term $p_{12}p_{23}/\mu_\pi$ is included reveals that its contribution to the scattering length amounts to 30–40 %.

2.11 Splitting of the Three-Particle Wave Function into "Internal" and "External" Parts

We can separate the three-particle wave function into "internal" and "external" parts. Such a separation can be performed without an explicit introduction of the radius separating the external region from the internal one, as is usually done in the R-matrix theory.

The idea of such a division [54] is based on the Noyes–Kowalski representation [2] (1.32) for the pair t-matrix

$$t(k, k', Z) = f(k, Z)\tau(Z)f(k', Z) + (k^2 - Z)(\pi/2)r(k, k', Z) . \qquad (2.242)$$

Upon substitution of this expression into the Faddeev equations, the second term leads to kernels of equations that contain no free Green's function. In other words, at a large distance these kernels yield no contribution to the three-particle wave function.

Let us illustrate this procedure by utilizing the example of scattering of a spinless particle on a bound pair. For simplicity, we shall restrict ourselves everywhere to the case of zero angular momenta. In this case the radial wave function of three particles in the configuration space is

$$\psi(x, y; Z) = \psi_\kappa(x) \sin(Z + \kappa^2)^{1/2} y$$
$$+ \frac{1}{\pi^2} \int_0^\infty dp^2 \int_0^\infty dq^2 \frac{K(p, q, Z)}{p^2 + q^2 - Z} \sin px \sin qy , \qquad (2.243)$$

where x and y are the Jacobi variables

$$x = \left(\frac{2m_2 m_3}{m_2 + m_3}\right)^{1/2} (r_2 - r_3) ;$$

$$y = \left[\frac{2m_1(m_2 + m_3)}{m_1 + m_2 + m_3}\right]^{1/2} \left(r_1 - \frac{m_2 r_2 + m_3 r_3}{m_2 + m_3}\right) .$$

The function K is related to the Faddeev components K_s:

$$K(p, q, Z) = K_1(p, q, Z) + \sum_{s=2,3} \int_{a_s}^{b_s} d\varphi K_s(P \cos \varphi, P \sin \varphi) , \qquad (2.244)$$

where

$$a_s = |\xi - \mu_{1s}| ; \quad b_s = \min(\xi - \mu_{1s}, \pi - \xi - \mu_{1s}) , \quad \xi = \arctan(q/p) ;$$
$$p^2 = p^2 + q^2 ; \quad \cos^2 \mu_{ss'} = m_1 m_s/(m_1 + m_{s'})(m_s + m_{s'}) .$$

The Faddeev equations for the components K_s are

$$K_s = \frac{1}{2\pi_q} \sum_{s' \neq s} \int_0^\infty \frac{dq^2}{\sin 2\mu_{ss'}} \int_{(p_{ss'}^-)^2}^{(p_{ss'}^+)^2} dp'^2 \frac{t_s(p, \bar{p}; Z - q^2)}{p'^2 + q'^2 - Z}$$

$$[\delta_{1s'}\psi_\kappa(p')\delta(q' - \sqrt{Z + \kappa^2}) + K_{s'}(p', q')Z] ; \left.\begin{array}{c} \\ \\ \\ \end{array}\right\} \qquad (2.245)$$
$$p_{ss'}^\pm = p' \cot \mu_{ss'} \pm q \csc \mu_{ss'} ; \quad \bar{p}^2 = p'^2 + q'^2 - q^2 ,$$
$$\psi_\kappa(p') = \varphi_\kappa(p')/(p'^2 + \kappa^2) ,$$

where φ_κ is the wave function of the bound pair.

Now we write the Noyes–Kowalski representation for t_s:

$$t_s(p, \bar{p}; Z - q^2) = f^s(p, Z - q^2)\tau_s(Z - q^2)f^s(\bar{p}, Z - q^2)$$
$$+ (\pi/2)(p^2 + q^2 - Z)r_s(p, \bar{p}, Z - q^2) . \tag{2.246}$$

Substituting (2.246) into the Faddeev equations (2.245) for K_s we obtain

$$K_s(p, q, Z) = f^s(p, Z - q^2)\tau_s(Z - q^2)H_s(q, Z)$$
$$+ (p^2 + q^2 - Z)I_s(p, q, Z) . \tag{2.247}$$

Functions H_s and I_s satisfy the set of coupled equations

$$H_s(q, Z) = \frac{1}{2\pi q} \sum_{s' \neq s} \int_0^\infty \frac{dq'^2}{\sin 2\mu_{ss'}} \int_{(p_{ss'}^-)^2}^{(p_{ss'}^+)^2} dp'^2 \frac{f^s(\bar{p}, Z - q^2)}{p'^2 + q'^2 - Z}$$

$$\times [\delta_{1s'}\psi_\kappa(p')\delta(q' - \sqrt{Z + \kappa^2} + K_{s'}(p', q', Z)] ;$$

$$I_s(p, q, Z) = \frac{1}{4q} \sum_{s' \neq s} \int_0^\infty \frac{dq'^2}{\sin 2\mu_{ss'}} \int_{(p_{ss'}^-)^2}^{(p_{ss'}^+)^2} dp'^2 \frac{r_s(p, \bar{p}, Z - q^2}{p'^2 + q'^2 - Z}$$

$$\times [\delta_{1s'}\psi_\kappa(p')\delta(q' - \sqrt{Z + \kappa^2}) + K_{s'}(p', q', Z)] . \tag{2.248}$$

From (2.247, 248) it follows that the two-dimensional part K_s, i.e., function $I_s(p, q, Z)$, is responsible for the internal region, while the one-dimensional $H_s(q, Z)$ is responsible for the external region. On the basis of such a separation one can develop an approximate procedure for solving the two-dimensional Faddeev equations by expanding the function within the internal region into finite series over hyperspherical functions.

It is well known that for localized functions, series over hyperspherical functions converge quite rapidly.

The second question concerns the explicit demonstration of the advantages of the Faddeev equations over the Schrödinger equation, even for bound states [55–57]. By using a two-center Coulomb potential we can verify how important it is to split the wave function of the system into Faddeev components.

The two-center problem with Coulomb interaction is usually solved by transforming the Schrödinger equation to spheroidal coordinates, in which the variables separate. We shall not convert to these variables but shall apply the method of coupled channels to the solution of equations of the Faddeev type, i.e., of the equations for the components of the wave function which have physical asymptotics.

Let us first consider a system consisting of a single light negative particle (1) and two heavy positive particles (2 and 3). We postulate the following set of equations describing this system:

$$(E - H_0 - V_{12})\psi_1 = a_{11}\psi_1 + a_{12}\psi_2 ;$$
$$(E - H_0 - V_{31})\psi_2 = a_{21}\psi_1 + a_{22}\psi_2 . \tag{2.249}$$

Here $\psi = \psi_1 + \psi_2$ is the wave function of the system; r_1 is the relative distance between particles 1 and 2; ϱ is the relative distance between particles 2 and 3; $r_3 = \varrho - r_1$;

$$V_{12}(r_1) = -1/r_1 ; \quad V_{23}(\varrho) = 1/\varrho ; \quad V_{13}(r_3) = -1/r_3 .$$

We shall require the set of equations (2.249) to be equivalent to the Schrödinger equation. This leads to the conditions

$$a_{11} + a_{21} = V_{13} + V_{23} ; \quad a_{12} + a_{22} = V_{12} + V_{23} . \tag{2.250}$$

We shall search for the solution of (2.249) in the form of expansions

$$\psi_1 = \sum_n F_n(\varrho)\varphi_n(r_1) ; \quad \psi_2 = \sum_n F_{\bar{n}}(\varrho)\varphi_{\bar{n}}(r_3) , \tag{2.251}$$

where $\varphi_n(r_1)$ and $\varphi_{\bar{n}}(r_3)$ are the hydrogenlike eigenfunctions of the bound system of the light particle 1 and nucleus 2 and of particle 1 and nucleus 3, respectively. In (2.251), besides summation, integration over the continuous spectrum is also implied.

As can be seen from (2.250), these relations do not completely fix the coefficients a_{ij}. Assuming the effective interaction energy of the heavy particles to exhibit correct asymptotic behavior as ϱ tends to infinity, we arrive at the condition $a_{11} = a_{22} = 0$. As a result, (2.249) assume the form

$$(E - H_0 - V_{12})\psi_1 = (V_{12} + V_{23})\psi_2 ;$$
$$(E - H_0 - V_{13})\psi_2 = (V_{13} + V_{23})\psi_1 . \tag{2.252}$$

Before proceeding to solve them, we shall make some remarks concerning the uniqueness of the solutions. Lately, the possibility of the existence of the so-called ghost solutions of equations such as (2.252) has been raised. Ghost solutions are understood to be such solutions ψ_1 and ψ_2 which satisfy the condition

$$\psi_1 + \psi_2 = \psi = 0 . \tag{2.253}$$

Here $\psi_1 \neq 0$ and $\psi_2 \neq 0$. Clearly, in the case of the Faddeev equations, owing to the definition of the Faddeev components $\psi_i = G_0 V_i \psi$, condition (2.253) cannot be fulfilled and no ghost solutions are present.

We transform (2.252) to

$$\psi_1 = G_0(V_{12}\psi + V_{23}\psi_2) ; \quad \psi_2 = G_0(V_{13}\psi + V_{23}\psi_1) . \tag{2.254}$$

Combining (2.254) for the components ψ_1 and ψ_2 one can obtain formal expressions in terms of the wave function:

$$\psi_1 = W_1\psi ; \quad \psi_2 = W_2\psi , \tag{2.255}$$

where the operators W_1 and W_2 are given in explicity. Further, it is readily verified that these operators satisfy the relation

$$W_1 + W_2 = 1 . \tag{2.256}$$

Thus, if we have found nonzero solutions of (2.252), then because of (2.256), condition (2.253) will not hold, i.e., ghost solutions are absent in (2.252). In the above reasoning we bore in mind the exact solutions ψ_1 and ψ_2 and the exact operators $W_{1,2}$. If, on the other hand, as a result of an approximate solution of the equations, we find approximate operators $\tilde{W}_1$ and $\tilde{W}_2$ with the property $\tilde{W}_1 + \tilde{W}_2 \ll 1$, then these approximate solutions must be immediately discarded, because of their remoteness from the exact solutions which lead to (2.256).

We can now continue proceed with the solution of (2.252). We shall deal with the problem of two centers; in this case the equations

$$[E - h_1(\mathbf{r}_1)]\psi_1 = [V_{12}(\mathbf{r}_1) + V_{23}(\varrho)]\psi_2 ;$$
$$[E - h_2(\mathbf{r}_3)]\psi_2 = [V_{13}(\mathbf{r}_3) + V_{23}(\varrho)]\psi_1 , \tag{2.257}$$

where

$$h_1 = -\frac{1}{2m_1}\Delta r_1 + V_{12}(\mathbf{r}_1) ; \quad h_2 = -\frac{1}{2m_2}\Delta r_3 + V_{13}(\mathbf{r}_3) ; \tag{2.258}$$

m_1 and m_2 are the reduced masses of particles 1, 2 and 1, 3, respectively.

Substituting (2.251) into (2.257) and passing to the total angular momentum representation we obtain the set of equations

$$(E - E_m)f_m^{LM} = \sum_{\bar{n}} V_{m\bar{n}}^{(1)} f_{\bar{n}}^{LM} ;$$
$$(E - E_{\bar{m}})f_{\bar{m}}^{LM} = \sum_{n} V_{\bar{m}n}^{(2)} f_n^{LM} . \tag{2.259}$$

In (2.259) the following notation is adopted: $m \equiv (m, I_r, I_\varrho)$; $\bar{m} \equiv (\bar{m}, \bar{I}_r, \bar{I}_\varrho)$; $n \equiv (n, I_r', I_\varrho')$; $\bar{n} \equiv (\bar{n}, \bar{I}_r', \bar{I}_\varrho')$; inside the parentheses there are the principal quantum number and the values of the respective angular momenta

$$V_{m\bar{n}}^{(1)}(\varrho) = \sum_{\bar{\mu}_r, \bar{\mu}_\varrho', \mu_r, \mu_\varrho} (I_r\mu_r I_\varrho\mu_\varrho|LM)(I_r\bar{\mu}_r \bar{I}_\varrho'\bar{\mu}_\varrho'|LM)$$

$$\times \int Y_{I_\varrho\mu_\varrho}^*(\hat{\varrho})Y_{I_r\mu_r}^*(\hat{r}_1)R_{mI_r}(r_1(V_{12} + V_{23})R_{\bar{n}I_r}(r_3)$$

$$\times Y_{I_r\mu_r}(\hat{r}_3)Y_{\bar{I}_{\varrho'},\bar{\mu}_\varrho'}(\hat{\varrho})dr, d\Omega\varrho ;$$

$$V_{mn}^{(2)}(\varrho) = \sum_{\mu_r, \mu_\varrho', \mu_r', \mu_\varrho} (I_r\bar{\mu}_r \bar{I}_\varrho\bar{\mu}_\varrho|LM)(I_r\mu_r I_\varrho'\mu_\varrho'|LM)$$

$$\times \int Y_{\bar{I}_\varrho\bar{\mu}_\varrho}^*(\hat{\varrho})Y_{\bar{I}_r\bar{\mu}_r}^*(\hat{r}_3)R_{\bar{m}\bar{I}_r}(r_3(V_{13} + V_{23})R_{nI_r}(r_1)$$

$$\times Y_{I_r\mu_r}(\hat{r}_1)Y_{I_\varrho',\mu_\varrho'}(\hat{\varrho})dr_3d\Omega\varrho ;$$

To solve (2.259) we shall separate with respect to indices, i.e., we shall assume

$$V_{mn} \approx V_{mn}^N = \sum_{ij}^N V_{mi} d_{ij}^{-1} V_{jn} \ . \tag{2.260}$$

Limiting the consideration to the case of $N = 1$, (2.259) may be written as

$$\left.\begin{array}{l} V_{\bar{1}1}^{(2)} F_1 = \sum_m V_{\bar{1}m}^{(2)} (E - E_m)^{-1} V_{m\bar{1}}^{(1)} F_2 \ ; \\ V_{1\bar{1}}^{(1)} F_2 = \sum_{\bar{m}} V_{1\bar{m}}^{(1)} (E - E_m)^{-1} V_{\bar{m}1}^{(2)} F_1 \ . \end{array}\right\} \tag{2.261}$$

Since the set of equations (2.261) is algebraic, we readily find the dependence of the energy E upon the distance between particles 2 and 3, i.e, the effective potential for such atomic and mesic-atomic systems as $pp\mu$, $dT\mu$, $^3\mathrm{He}d\mu$, and others.

For an illustration of the effectiveness of (2.261) we shall consider the spectrum of the $dd\mu$ system in the state with $L = 0$. Since in this case the masses of the heavy particles are identical, (2.261) reduces to a single equation

$$V_{\bar{1}1}^{(2)} F = \sum_m V_{\bar{1}m}^{(2)} (E - E_m)^{-1} V_{m\bar{1}}^{(1)} F \ . \tag{2.262}$$

Taking into account only the ground state of the $d\mu$ mesic atom one can find the effective potential of the $dd\mu$ system:

$$E(\varrho) = \left(\frac{3}{2} m^2 \varrho + \frac{1}{\varrho} \right) \exp(-m\varrho) \ . \tag{2.263}$$

Here m is the reduced mass of d and μ (the energy is referenced to the ground level of the $d\mu$ atom). The potential (2.263) can be approximated to good accuracy by the Morse potential:

$$V_\mu(\varrho) = D \{ \exp[-2\alpha(\varrho - R_0)] - 2\exp[-\alpha(\varrho - R_0)] \}$$

with the parameters $R_0 = 2.19 \ a.u.$; $D = 0.107 \ a.u.$; $\alpha = 0.72 \ a.u..$ This leads to the energies of the ground and excited states: $E(\nu = 0) = -327 \, \mathrm{eV}$, $E(\nu = 1) = -28 \, \mathrm{eV}$. These energy levels are quite close to the ones obtained from cumbersome calculations with a large number of terms taken into account in the expansion of the three-particle function in the adiabatic basis [58].

3. The Four-Body Problem

In this chapter we would like to first make some general remarks and then, as a preparatory step, obtain the Faddeev equations in the differential form. Then we shall proceed to the derivation of the Yakubovsky equations in the differential and integral forms [59, 61]. We shall also obtain the AGS four-particle integral equations [60]. The approximate four-particle equations for the low-energy elastic scattering amplitudes will be discussed. The effectiveness of these equations will be demonstrated by two examples: $\pi^{\pm}$-^{3}He scattering and n-^{3}He scattering. On the basis of the approximations developed for four-particle systems, we will consider the six-body problem in elastic NN scattering within the framework of the *nonrelativistic quark model*.

The information existing on four-particle systems is significantly richer and more diverse than that on three-particle systems. Moreover, in such systems binary reactions , which are totally impossible in three-particle systems, may occur, for example: n+^{3}He $\rightarrow$ ^{3}H+p; π^-+^{3}He $\rightarrow$ ^{3}H+π^0. In addition, the radiative capture of thermal neutrons by ^{3}He, i.e., the reaction n+^{3}He $\rightarrow$ ^{4}He+γ, is much more sensitive to the contribution of the so-called exchange currents than the similar reaction in a three-particle system. In extracting detailed information on the structure of exchange meson currents, it is necessary to determine accurately of the wave functions of the four-particle systems ^{4}He and n+^{3}He on the basis of the same four-nucleon Hamiltonian.

Theoretical calculations and experimental values of the binding energy of the ^{4}He nucleus do not agree and there is no explanation for the behavior of the form factor of this nucleus. At the same time, practically no theoretical analysis of the four-particle data based on the solution of the exact four-particle Yakubovsky equations exist. There are, however, various calculations based on the AGS equations [60, P6, P7]. The Yakubovsky calculations that are available include for n^{3}He(^{3}H) – scattering lengths and of the amplitudes of certain processes in the K-matrix approximation [62], as well as the very rough calculation of the binding energies of the Λ^4He and Λ^4H nuclei [63].

In the computation of bound states of the four-particle systems, in particular, of the ^{4}He nucleus, significant progress has been achieved by the application of the method of hyperspherical functions and its various modifications [64]. Application of this method has made it possible to obtain converging results on the binding energy for practically all the existing "realistic" potentials. They all, however, yield a binding energy for the α-particle which is too small by 4–5 MeV.

3.1 The Integral and Differential Yakubovsky Equation

We first note that a rearrangement of the four-particle equations involving only the introduction of the Faddeev components

$$\psi\alpha' = -G_0 V_\alpha \psi \, , \tag{3.1}$$

is insufficient.

Fig. 3.1. Disconnected diagrams of a three-particle system: two particles interact through the potential V_α, the third one passes without undergoing interaction

We recall that the introduction of these components into the three-particle case led to the successful elimination of terms corresponding to disconnected diagrams of the type presented in Fig. 3.1 in the kernels of the equations.

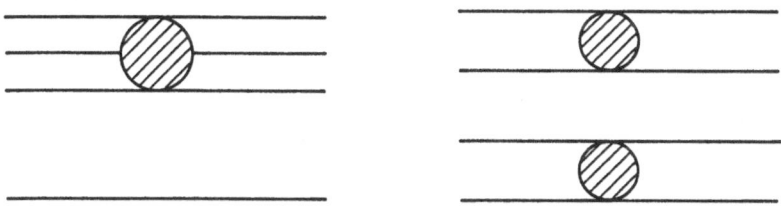

Fig. 3.2. Disconnected diagram of a four-particle system: three particles interact with each other, while the fourth particle passes without interaction

Fig. 3.3. Disconnected diagram of a four-particle system: the interaction differs from zero only in pair subsystems, the subsystems themselves move independently

In the four-particle situation there arise new disconnected diagrams (Figs. 3.2, 3) which must also be eliminated, since they lead to the appearance of the delta-function of the corresponding momenta.

Thus, we shall first repeat the derivation of the Faddeev equations. We have the three-particle Hamiltonian

$$H = H_0 + \sum_\alpha V_\alpha \, , \quad \alpha = \{12, 23, 13\} \, . \tag{3.2}$$

We shall rewrite the Schrödinger equation in as

$$(H_0 - E)\psi = -V\psi \tag{3.3}$$

or

$$\psi = -G_0(E)V\psi .$$ (3.4)

Here, for simplicity we restrict ourselves to the homogeneous equation (3.4). The Faddeev component defined by

$$\psi_\alpha = -G_0(E)V_\alpha\psi ,$$ (3.5)

obviously satisfies the relation

$$\psi = \sum_\alpha \psi_\alpha .$$ (3.6)

Relation (3.5) may be rewritten in the differential form

$$(H_0 - E)\psi_\alpha = -V_\alpha\psi .$$ (3.7)

Substituting (3.6) into the right-hand side of (3.7) and transferring the "diagonal" terms to the left-hand side, we obtain the differential Faddeev equations

$$(H_0 + V_\alpha - E)\psi_\alpha = -V_\alpha \sum_{\beta \neq \alpha} \psi_\beta .$$ (3.8)

Inverting the operator in the left-hand side of (3.8) we obtain the integral Faddeev equations:

$$\psi_\alpha = -G_\alpha(E)V_\alpha \sum_{\beta \neq \alpha} \psi_\beta .$$ (3.9)

The differential Faddeev equations (3.8) must, naturally, be supplemented with boundary conditions, which in this case are

$$\psi_\alpha^{(\beta)}(\boldsymbol{x}, \boldsymbol{p}_\beta) = \psi_\alpha(\boldsymbol{x}_\alpha)\exp(\mathrm{i}\boldsymbol{p}_\alpha\boldsymbol{y}_\alpha)\delta_{\alpha\beta}$$
$$+ \psi_\alpha(\boldsymbol{x}_\alpha)U_{\alpha\beta}(\boldsymbol{y}_\alpha\boldsymbol{p}_\beta) + U_{0\beta}(\boldsymbol{x}\boldsymbol{p}_\beta) ,$$ (3.10)

where $U_{\alpha\beta}(\boldsymbol{y}_\alpha, \boldsymbol{p}_\beta) \sim A_{\alpha\beta}(\hat{y}_\alpha, \boldsymbol{p}_\beta)(l/y_\alpha)\exp[\mathrm{i}\sqrt{E + \varepsilon_\alpha y_\alpha}]$; $A_{\alpha\beta}$ are the elastic scattering and exchange amplitudes; $U_{0\beta}(\boldsymbol{x}, \boldsymbol{p}_\beta) \sim A_{0\beta}(\boldsymbol{x}, \boldsymbol{p}_\beta)\exp[\mathrm{i}\sqrt{E}x]/x^{5/2}$; $A_{0\beta}$ is the breakup amplitude.

Here we have used the notation for the Jacobi variables shown in Fig. 3.4, $\boldsymbol{x} \sim \{\boldsymbol{x}_\alpha, \boldsymbol{y}_\alpha\}$.

We shall now treat the case of four particles by following the simple derivation of S.L. Yakovliev. Here various particle partitions are feasible, such as depicted in Figs. 3.1–3, therefore, we shall make use of such concepts as partitions and chains of partitions. We shall call the distribution of n particles into a groups the partition a. Examples of partitions of four particles into two groups are $(ijk)l$ and $(ij)(kl)$. Within these groups only those particles whose indices occur inside the parentheses interact. In the case of two partitions a and α the symbol (a, α)

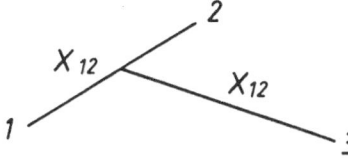

Fig. 3.4. Jacobi variables in a system consisting of three particles: x_{12} describes the position of particles 1 and 2 with respect to each other, y_{12} describes the relative position of particle 3 and the center of mass of particles 1 and 2

will be called a chain of partitions. The symbol $a \supset \alpha$ indicates that the partition α is obtained from the partition a by partitioning one or more of its subsystems.

We again begin by introducing the Faddeev components of the wave function,

$$\psi_\alpha = -G_0(E)V_\alpha\psi . \tag{3.11}$$

As in the three-particle case, the component ψ_α satisfies the equations

$$(H_0 + V_\alpha - E)\psi_\alpha = -V_\alpha \sum_{\beta \neq \alpha} \psi_\beta \tag{3.12}$$

or in the integral form

$$\psi_\alpha = -G_\alpha(E)V_\alpha \sum_{\beta \neq \alpha} \psi_\beta . \tag{3.13}$$

Equations (3.13) are still not Fredholm equations. For the ultimate formulation of these equations we introduce, with the aid of the chain (α, a), components for each function ψ_α (by exact analogy with the three-particle case):

$$\psi_{\alpha a} = -G_\alpha(E)V_\alpha \sum_{(\beta \neq \alpha) \subset a} \psi_\beta \tag{3.14}$$

or the same thing in the differential form:

$$(H_0 + V_\alpha - E)\psi_{\alpha a} = -V_\alpha \sum_{(\beta \neq \alpha) \subset a} \psi_\beta . \tag{3.15}$$

Thus, if the Faddeev component ψ_α "proportional" to the particle interaction in the pair α is specified by the index of the last interacting pair of particles, then its component $\psi_{\alpha a}$ is also identified by the partition (in this case, a) from which the pair α arose.

Let us continue with the derivation of the equations. It can be shown [58] that the following summation rule holds:

$$\sum_{a \supset \alpha} \sum_{(\beta \neq \alpha) \subset a} = \sum_{\beta \neq \alpha} . \tag{3.16}$$

Applying this rule to both the sides of (3.14) we obtain the relation

$$\psi_\alpha = \sum_a \psi_{\alpha a} . \tag{3.17}$$

Substituting (3.17) into (3.15) we immediately get

$$(H_0 + V_\alpha - E)\psi_{\alpha a} = -V_\alpha \sum_{(\beta \neq \alpha) \subset a} \sum_b \psi_{\beta b} . \tag{3.18}$$

Following the Faddeev idea as in the three-particle case, we single out in the right-hand side of (3.18) the "diagonal" term, i.e., the term with $b = a$, and transferring it to the left-hand side we obtain the differential Yakubovsky equations [60]

$$(H_0 + V_\alpha - E)\psi_{\alpha a} + V_\alpha \sum_{(\gamma \neq \alpha) \subset a} \psi_{\gamma a} = -V_\alpha \sum_{b \neq a\,(\beta \neq \alpha) \subset a} \sum \psi_{\beta b} . \tag{3.19}$$

We shall now transform (3.19) to its integral form. Clearly, to do so it is necessary to invert the operator acting on the components of the function in the left-hand side of (3.19). Unlike the three-particle case, this now represents a matrix operator. Obviously, the desired inverse operator $G^a_{\alpha\beta}$, by definition, must satisfy the set of equations

$$(H_0 + V_\alpha - E)G^a_{\alpha\beta} + V_\alpha \sum_{(\gamma \neq \alpha) \subset a} G^a_{\gamma\beta} = -I\delta_{\alpha\beta} , \tag{3.20}$$

where I is the unit operator. Now, (3.20) is just the Faddeev equation for the components of the Green function generated by the Hamiltonian

$$H_a = H_0 + \sum_{\alpha \subset a} V_\alpha . \tag{3.21}$$

The Faddeev equations (3.20) in the integral form are more familiar:

$$G^a_{\alpha\beta} = -G_\alpha \delta_{\alpha\beta} - G_\alpha V_\alpha \sum_{(\gamma \neq \alpha) \subset a} G^a_{\gamma\beta} . \tag{3.22}$$

Inverting (3.19) with the aid of (3.20) we obtain the integral Yakubovsky equations for the wave function:

$$\psi_{\alpha a} = - \sum_{b \neq a\,(\gamma \neq \beta) \subset a} \sum G^a_{\alpha\gamma} V_\alpha \psi_{\beta b} . \tag{3.23}$$

We shall now impose the boundary conditions at infinity onto the differential Yakubovsky equations (3.19). To this end we introduce the Jacobi variables for four particles (Fig. 3.5). For convenience we shall assume

$$X \equiv \{\boldsymbol{x}_\alpha, \boldsymbol{y}_{\alpha a}, \boldsymbol{y}_a\} ; \quad \boldsymbol{x}_a \equiv \{\boldsymbol{x}_\alpha, \boldsymbol{y}_{\alpha a}\} ; \quad \boldsymbol{y}_\alpha \equiv \{\boldsymbol{y}_{\alpha a}, \boldsymbol{y}_a\} .$$

Then for the wave function we obtain the asymptotic conditions

$$\psi^{(b)}_{\alpha a}(X, \boldsymbol{p}_b) \approx \psi^\alpha_a(\boldsymbol{x}_a) \exp(i\boldsymbol{p}_a \boldsymbol{y}_a)\delta_{ab} + \psi^\alpha_a(\boldsymbol{x}_a)U_{ab}(\boldsymbol{y}_a, \boldsymbol{p}_b)$$
$$+ \psi_\alpha(\boldsymbol{x}_\alpha)U^a_{ab}(\boldsymbol{y}_\alpha, \boldsymbol{p}_b) + U^{\alpha a}_{0b}(\boldsymbol{x}, \boldsymbol{p}_b) , \tag{3.24}$$

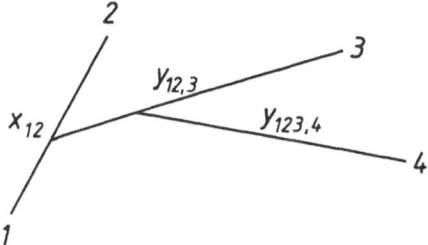

Fig. 3.5. Jacobi variables for a four-particle system

where the following notation is adopted:

$$(2 \to 2): \quad U_{ab}(\boldsymbol{y}_a, \boldsymbol{p}_b) \sim A_{ab}(\hat{y}_a, \boldsymbol{p}_b)\frac{\exp\{i\sqrt{E + \varepsilon_a}|\boldsymbol{y}_a|\}}{|\boldsymbol{y}_a|} \; ;$$

$$(2 \to 3): \quad U_{\alpha b}^a(\boldsymbol{y}_\alpha, \boldsymbol{p}_b) \sim A_{\alpha b}^a(\hat{y}_\alpha, \boldsymbol{p}_b)\frac{\exp\{i\sqrt{E + \varepsilon_\alpha}|\boldsymbol{y}_\alpha|\}}{|\boldsymbol{y}_\alpha|^{5/2}} \; ; \qquad (3.25)$$

$$(2 \to 4): \quad U_{0b}^{\alpha a}(\boldsymbol{x}, \boldsymbol{p}_b) \sim A_{0b}^{\alpha a}(\hat{x}, \boldsymbol{p}_b)\frac{\exp\{i\sqrt{E}|\boldsymbol{x}|\}}{|\boldsymbol{x}|^4} \; .$$

As is seen from the structure of the differential equations (3.19) and their boundary conditions, for the solution of scattering problems in four-particle systems only the knowledge of the wave functions of the bound states of the subsystems is necessary. On the other hand, to solve the integral Yakubovsky equations (3.23) one must have the Green functions or the scattering amplitudes for the subsystem throughout the whole range of values of the variables. Therefore, it is absolutely clear that the solution of the differential Yakubovsky equations represents a significantly simpler problem.

3.2 Four-Particle AGS Equations

The AGS form of the four-particle equations is widely used [62]. We recall that the AGS equations are equations for the transition operators, but not for the wave functions. Just as in the case of the Yakubovsky equations, the derivation of the AGS equations is based on the dea of a double subsequent rearrangement of the kernels of the Lippmann–Schwinger equations, as a result of which they become equations of the Fredholm type.

We shall need some relations concerning the technique of the AGS equations from the second lecture; for convenience we shall review them here. The equations for the transition operators $U_{\beta\alpha}$ are

$$U_{\beta\alpha} = (1 - \delta_{\beta\alpha})G_0^{-1} + \sum_{\gamma \neq \beta} T_\gamma G_0 U_{\gamma\alpha} \; . \qquad (3.26)$$

This set of equations may, as already noted, be written formally in the form of the Lippmann–Schwinger equation for the T-matrix:

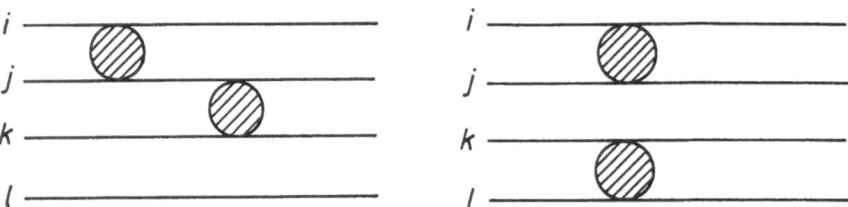

Fig. 3.6. Disconnected four-particle diagram: the l-th particle is free

Fig. 3.7. Disconnected four-particle diagram: the ij pair of particles does not interact with the kl pair

$$T = V + V\mathcal{Y}_0 T \qquad (3.27)$$

where

$$T_{\beta\alpha} \equiv U_{\beta\alpha} ; \quad V_{\beta\alpha} \equiv (1 - \delta_{\beta\alpha})G_0^{-1} ; \quad \mathcal{Y}_{0,\beta\alpha} \equiv \delta_{\beta\alpha}G_0 T_\alpha G_0 . \qquad (3.28)$$

If we now consider that the matrix indices in (3.26, 27) can vary between 1 and 6 instead of 1 and 3, i.e., if we thus turn to the four-particle system, we find that these equations have no unique solutions, since their kernels are not integrable in quadratures. Indeed, the expression $V\mathcal{G}_0$, as it follows from (3.28), involves an interaction between two of the four particles. This leads to a delta-function with respect to the variables of the remaining particles arising precisely as in the kernel of the three-particle Lippmann–Schwinger equation. In other words, in the four-body problem the equation

$$\mathcal{Y} = \mathcal{Y}_0 + \mathcal{Y}_0 V \mathcal{Y} \qquad (3.29)$$

plays approximately the same role as the Green function $G = G_0 + G_0 \sum_{\gamma=1}^{3} V_\gamma G$ in the three-body problem. Hence, it is necessary once more to apply the Faddeev procedure of "Fredholmization" of the kernels of (3.27). To this end we introduce the cluster indices ϱ, σ, and τ which we shall make to denote feasible partitions in the four-body system. For example, the partition $\varrho = (ijk, l)$ corresponds to the diagram in Fig. 3.6, while the partition $\varrho = (ij, kl)$ corresponds to the diagram in Fig. 3.7. With this notation, we rewrite (3.29) in the form

$$\mathcal{Y} = \mathcal{Y}_0 + \mathcal{Y}_0 \sum_\varrho V^\varrho \mathcal{Y}_0 , \qquad (3.30)$$

where V^ϱ represents the part of V acting only the cluster ϱ, i.e., by analogy with (3.28), we have

$$V_{\beta\alpha}^\varrho = (1 - \delta_{\beta\alpha})G_0^{-1} , \qquad (3.31)$$

if $\beta, \alpha \in \varrho$, i.e., if the pairs $\alpha = (i, j)$ and $\beta = (i', j')$ occur in the sets of indices characterizing the given ϱ. Otherwise, $V_{\beta\alpha}^\varrho = 0$.

Thus, continuing the analogy with the three-particle case we introduce an analogue of the Faddeev operator $M_{\beta\alpha}$:

$$\mathcal{M}^{\sigma\varrho} = \delta_{\sigma\varrho}\mathcal{V}^{\varrho} + \mathcal{V}^{\sigma}\mathcal{Y}\mathcal{V}^{\varrho} . \tag{3.32}$$

Further, in the same way as in Chap. 2 equations for the operators $M_{\beta\alpha}$ were obtained, we shall obtain equations for the operators $\mathcal{M}^{\sigma\varrho}$:

$$(1 - \mathcal{V}^{\sigma}\mathcal{Y}_0)\mathcal{M}^{\sigma\varrho} = \delta_{\sigma\varrho}\mathcal{V}^{\varrho} + \mathcal{V}^{\sigma}\mathcal{Y}_0 \sum_{\tau \neq \sigma} \mathcal{M}^{\tau\varrho} . \tag{3.33}$$

Inverting the operator occurring in the left-hand side of (3.33) in the parentheses we obtain the four-particle analogue of the Faddeev equations:

$$\mathcal{M}^{\sigma\varrho} = \delta_{\sigma\varrho}\mathcal{T}^{\varrho} + \mathcal{T}^{\sigma}\mathcal{Y}_0 \sum_{\tau \neq \sigma} \mathcal{M}^{\sigma\varrho} , \tag{3.34}$$

where the three-particle operator $\mathcal{T}^{\varrho}$ is defined as the solution of (3.27) with the potential $\mathcal{V}^{\varrho}$, i.e.,

$$\mathcal{T}^{\varrho} = \mathcal{V}^{\varrho} + \mathcal{V}^{\varrho}\mathcal{Y}_0\mathcal{T}^{\varrho} . \tag{3.35}$$

The same arguments which allow us in the three-particle case to transform from the operators $M_{\alpha\beta}$ to the operators $U_{\alpha\beta}$ permit transition from (3.34) to the equations for the transition operators of the four-particle system:

$$U^{\sigma\varrho} = (1 - \delta_{\sigma\varrho})\mathcal{Y}_0^{-1} + \sum_{\tau \neq \sigma} \mathcal{T}^{\tau}\mathcal{Y}_0 U^{\tau\varrho} . \tag{3.36}$$

Equations (3.35) for the three-particle transition operator $\mathcal{T}^{\varrho}$ in terms of the adopted notation (3.28) have the form

$$U^{\varrho}_{\beta\alpha} = (1 - \delta_{\beta\alpha})G_0^{-1} + \sum_{\substack{\gamma \neq \beta \\ \gamma \in \varrho}} T_{\gamma}G_0 U^{\varrho}_{\gamma\alpha} . \tag{3.37}$$

The equations for the four-particle operators $\mathcal{M}^{\sigma\varrho}$ and $U^{\sigma\varrho}$ are written explicitly as

$$M^{\sigma\varrho}_{\beta\alpha} = \delta_{\sigma\varrho}U^{\varrho}_{\beta\alpha} + \sum_{\gamma} U^{\sigma}_{\beta\gamma}G_0 T_{\gamma}G_0 \sum_{\tau \neq \sigma} M^{\tau\varrho}_{\gamma\alpha} ; \tag{3.38}$$

$$U^{\sigma\varrho}_{\beta\alpha} = (1 - \delta_{\sigma\varrho})\delta_{\beta\alpha}G_0^{-1}T_{\alpha}^{-1}G_0^{-1} + \sum_{\substack{\tau \neq \sigma \\ \gamma}} U^{\tau}_{\beta\gamma}G_0 T_{\gamma}G_0 U^{\tau\varrho}_{\gamma\alpha} . \tag{3.39}$$

Like the integral Yakubovsky equations, the four-particle AGS equations (3.38, 39) contain as kernels the transition operators for the two- and three-particle subsystems T_{γ} and $U^{\beta}_{\tau\gamma}$, respectively.

In solving the complex sets of integral equations in the Yakubovsky form (3.23) or in the AGS form (3.39), which in the general case represent sets of nine-dimensional equations, one takes, naturally, advantage of an approximation for the kernels of these equations. Usually, these approximations use various separable, i.e. finite-dimensional, representations of the two- and the three-particle amplitudes determining the kernels of the four-particle equations [60]. An example is the widely applied expansion over the corresponding functions of the Hilbert–Schmidt problem [65]. However, construction of such functions and of the respective eigenvalues, even in the two-body problem, can be performed for the majority of known short-range potentials only numerically. The construction of three-particle Hilbert–Schmidt functions represents an even more difficult task [66].

Thus, the solution of the four-particle integral equations necessitates using various approximations and controlling the convergence of these approximations. The problem becomes even more complicated if one wishes to deal with "realistic" NN potentials. The use of these complex potentials leads to a catastrophic increase in the number of components in the four-particle wave function and to a corresponding increase in the dimensionality of the equations to be solved.

3.3 Approximate Four-Particle Equations for the π (3N) System

Let us attempt to clarify the physical reasons leading to equations of such a complicated structure and then, taking advantage of this understanding, we shall formulate new approximate four-particle equations, more convenient for practical computations.

As has been noted in Chap. 2, the physical meaning of the Faddeev components of the wave function of a three-particle system is that each component corresponds to a quite definite asymptotic behavior of the system. Since in a four-particle system the number of possible asymptotic states is higher than in the case of a three-particle system the number of components of the wave function in is also larger. Precisely this permits, in certain cases, one to consider the four-particle equations involving a smaller number of components in the wave function [67].

Consider the scattering of a particle on a bound complex of three particles at an incident energy that does not exceed the nearest breakup threshold of the three-particle target. In this case the asymptotics of the four-particle wave function is

$$\langle \boldsymbol{r}_{12}, \boldsymbol{r}_3, \varrho | \psi \rangle \equiv \psi(\boldsymbol{r}_{12}, \boldsymbol{r}_3, \varrho)$$

$$\approx \psi_0(\boldsymbol{r}_{12}, \boldsymbol{r}_3, \varrho) + T\chi(\boldsymbol{r}_{12}, \boldsymbol{r}_3)\frac{\exp(i\kappa\varrho)}{\varrho} ; \quad \varrho \to \infty . \quad (3.40)$$

$$\psi(\boldsymbol{r}_{12}, \boldsymbol{r}_3, \varrho) \to 0 , \quad |\boldsymbol{r}_{12}|, |\boldsymbol{r}_3| \to \infty ,$$

where $\psi_0 = \kappa e^{i\kappa\varrho}$ is the incident wave, and the Jacobi variables are presented in Fig. 3.8; $\kappa = \sqrt{2\mu E_0}$; μ is the reduced mass of the target and incident particle; E_0 is the kinetic energy of the incident particle; T is the sought elastic scattering

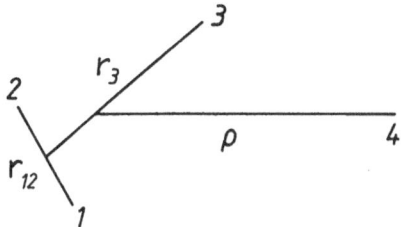

Fig. 3.8. "Tree-like" set of Jacobi variables for a four-particle system

amplitude, and $\langle r_{12}, r_3 | \chi \rangle \equiv \chi(r_{12}, r_3)$ is the wave function of the ground state of the target. Owing to the boundary conditions (3.40) the four-particle wave function satisfies the condition

$$\int |\psi(r_{12}, r_3, \varrho)|^2 dr_{12} dr_3 \leq M ,$$
(3.41)

i.e., it represents a function quadratically integrable over the coordinates of the nucleons. This is the key fact for all subsequent reasoning. We shall bear in mind a four-particle system consisting of a pion and the three-nucleon nucleus ^{3}He (^{3}H). (A more general case of four identical particles will be considered below.) Within the framework of the potential picture the Hamiltonian of this four-particle system has the form

$$H = h_0 + h_c + V_N + V_\pi ,$$
(3.42)

where

$$V_N = \sum_{i \neq j=1}^{3} V_N^{ji} ; \quad V_\pi = \sum_{i=1}^{3} V_{\pi N}^{i} ;$$

h_0 is the kinetic energy of the relative motion of the pion and the center of mass of the nucleus; h_c is the kinetic energy of the nucleons; V_N^{ij} is the interaction potential of the i-th and j-th nucleons in the target nucleus; $V_{\pi N}^{i}$ is the interaction potential of the pion and the i-th nucleon. On the basis of condition (3.41) one can approximate the part of the Hamiltonian (3.42) $h_c + V_N$ that depends only on the variables of the nucleons r_{12} and r_3 by an operator of finite rank. For example, one may assume

$$(h_c + V_N)|\psi\rangle \approx |\varphi\rangle\langle\varphi|h_c + V_N|\varphi\rangle\langle\varphi|\psi\rangle ,$$
(3.43)

where $\langle r_{12}, r_3 | \varphi \rangle \equiv \varphi(r_{12}, r_3)$ is a function with a finite norm. Clearly, if the function $\varphi(r_{12}, r_3)$ is an eigenfunction of the Hamiltonian of the target for example, $\varphi(r_{12}, r_3) = \chi(r_{12}, r_3)$, then $\langle\varphi|h_c + V_N|\varphi\rangle = \varepsilon$, where ε is the energy of the ground state of the three-nucleon system. The quantity $\langle\varphi|\psi\rangle$ represents a new unknown function, which now, however, depends only on the single variable ϱ. This makes it possible to obtain a simple one-dimensional equation for the elastic π^3He-scattering amplitude at a low pion energy. In a more complex case, when the system consists of several interacting subsystems (clusters), one can substitute the operators of a finite rank for the parts of the total Hamiltonian, depending only on variables of the bound subsystems, and thus reduce essentially and, at the same time, mathematically correctly, the dimensionality of the initial equations.

To make use of a device such as (3.43), we rewrite the Lippmann–Schwinger equation for the system $\pi + 3N$ singling out in it the Hamiltonian of the bound system in its explicit form. We introduce the Green functions:

$$G(E) = (H - E)^{-1} ; \quad G_c(E) = (h_0 + h_c + V_N - E)^{-1} ;$$
$$G_0(E) = (h_0 - E)^{-1} ; \quad G_{0\theta}(E) = (h_0 + h_c - E)^{-1} ;$$
$$G_\pi(E) = (h_0 + V_\pi - E)^{-1} . \tag{3.44}$$

The four-particle transition operator is

$$T = V_\pi - V_\pi G(E)V_\pi . \tag{3.45}$$

Further we shall need the auxiliary, also the four-particle, transition operator T^0 defined as

$$T^0 = V_\pi - V_\pi G_\pi(E)V_\pi = V_\pi - V_\pi G_0(E)T^0 . \tag{3.46}$$

Taking advantage of the equation for the total Green function

$$G(E) = G_\pi(E) - G_\pi(E)H_c G(E) , \tag{3.47}$$

where $H_c = h_c + V_N$, and of definitions (3.44, 46), we obtain the equation for the operator T in a form convenient for the approximation which we wish to perform:

$$T = T^0 + T^0 G_0(E)H_c G_c(E)T = T^0 + T^0[G_0(E) - G_c(E)]T . \tag{3.48}$$

Note that (3.46) represents a version of the equation for the amplitude describing pion scattering on three fixed centers. In the momentum representation this equation is

$$\langle k|\tau^0(r_{12}, r_3)|k'\rangle = f(k, k', r_{12}, r_3)\langle k|v_\pi|k'\rangle$$
$$- (2\pi)^{-3} \int dk'' f(k, k'', r_{12}, r_3)\langle k|v_\pi|k''\rangle \frac{\langle k''|\tau^0(r_{12}, r_3|k'\rangle}{k''^2/(2\mu) - E} , \tag{3.49}$$

where the following notation has been adopted:

$$\langle k, r_{12}, r_3|T^0|k', r'_{12}, r'_3\rangle = \delta(r_{12} - r'_{12})\delta(r_3 - r'_3)\langle k|\tau^0(r_{12}, r_3)|k'\rangle ;$$
$$f(k, k', r_{12}, r_3) = \exp\left[i(k' - k)\left(\frac{r_{12}}{2} + \frac{r_3}{3}\right)\right]$$
$$+ \exp[i(k' - k)(-r_{12}/2 + r_3/3)] \tag{3.50}$$
$$+ \exp\left[-i(k' - k)r_3\tfrac{2}{3}\right] ;$$

k and k' are the momenta describing the relative motion of the pion and the center of mass of the nucleus in the initial and final states. The amplitude τ^0 differs from the scattering amplitude on fixed centers, usually applied in the literature [17], first, in that the amplitude τ^0 from (3.49) is found at an energy equal to the total energy of the system, while the conventional scattering amplitude on fixed centers is determined at an energy equal to the kinetic energy of the incident particle. This difference leads, for example, to the amplitude τ^0, which plays the role of an effective "potential" in (3.48), being real below the breakup threshold and complex above the threshold. This seems quite justified from a physical point of view. Second, this amplitude describes the relative motion of the incident particle and the center of mass of the target, i.e., the target is not considered rigidly connected with the laboratory reference system, as is usually assumed to be in the model of fixed centers.

From the structure of the kernel of (3.49) it follows that it can be integrated analytically in the case of separable πN potentials v_π.

We now proceed with the approximation, namely the Hamiltonian of the nuclear subsystem H_c is approximated by an operator of a finite rank:

$$H_c \approx H_c^N = \sum_{\mu,\nu=1}^{N} |\chi_\mu\rangle \sigma_{\mu\nu} \langle\chi_\nu| \, . \tag{3.51}$$

Approximation (3.51) will be the sole one in obtaining the approximate four-particle equations. Writing H_c^N in the form of (3.51) actually represents various versions of substitution of the operator H_c^N of finite rank for the operator H_c, each differing from the other in the definition of the matrix $\sigma_{\mu\nu}$ and the state vectors $|\chi_\mu\rangle$. Bearing in mind determination of the $\pi^3 He$ elastic scattering amplitude, we shall assume that the functions $\chi_\mu(r_{12}, r_3) \equiv \langle r_{12}, r_3|\chi_\mu\rangle$ form an orthonormalized set of functions with a finite norm, χ_1 representing the wave function of the ground state of the bound three-particle system. In this case

$$\sigma_{\mu\nu} = \langle\chi_\mu|H_c|\chi_\nu\rangle \, , \quad \mu,\nu \neq 1 \, , \quad \sigma_{\mu 1} = \varepsilon\delta_{\mu 1} \, , \tag{3.52}$$

where ε is the binding energy of the three-nucleon nucleus. It is readily found that the Green function $G_c(E)$ in the approximation of (3.51) can be represented as

$$\langle k, r_{12}, r_3|G_c(E)|k' r_{12}', r_3'\rangle = (2\pi)^3 \delta(k - k')$$
$$\times [\delta(r_{12} - r_{12}')\delta(r_3 - r_3')(k^2/(2\mu) - E)^{-1}$$
$$- \sum_{\mu\nu} \chi_\mu(r_{12}, r_3)\Gamma_{\mu\nu}(k, E)\bar\chi_\nu(r_{12}', r_3')] \, . \tag{3.53}$$

Here

$$\Gamma_{\mu\nu}(k, E) = \sum_\lambda \frac{\sigma_{\mu\lambda}}{k^2/(2\mu) - E} \left[\left(\frac{k^2}{2\mu} - E + \hat{I}\hat\sigma\right)^{-1} \right]_{\lambda\nu} \, ,$$

where $I_{\mu\nu} = \langle\chi_\mu|\chi_\nu\rangle = \delta_{\mu\nu}$.

Thus, (3.51) permits transformation of the principal equation for the operator T, (3.48), to the form

$$\langle k, r_{12}, r_3 | T(E) | k', r'_{12}, r'_3 \rangle = \delta(r_{12} - r'_{12}) \delta(r_3 - r'_3) \langle k | \tau^0(r_{12}, r_3) | k' \rangle$$

$$+ \sum_{\mu\nu} \int \frac{dk''}{(2\pi)^3} dr''_{12} dr''_3 \langle k | \tau^0(r_{12}, r_3) | k'' \rangle$$

$$\times \chi_\mu(r_{12}, r_3) \Gamma_{\mu\nu}(k'', E) \bar\chi_\nu(r''_{12}, r''_3)$$

$$\times \langle k'', r''_{12}, r''_3 | T(E) | k', r'_{12}, r'_3 \rangle . \qquad (3.54)$$

Let us introduce new amplitudes

$$\langle k | t_{\mu\nu} | k' \rangle \equiv \int dr_{12} dr_3 dr'_{12} dr'_3 \bar\chi_\mu(r_{12}, r_3)$$

$$\times \langle k, r_{12}, r_3 | T(E) | k', r'_{12}, r'_3 \rangle \chi_\nu(r'_{12}, r'_3) ; \qquad (3.55)$$

$$\langle k | t^0_{\mu\nu} | k' \rangle \equiv \int dr_{12} dr_3 \bar\chi_\mu(r_{12}, r_3) \langle k | \tau^0(r_{12}, r_3) | k' \rangle \chi_\nu(r_{12}, r_3) .$$

Using these new amplitudes one can write the set of equations (3.54) as

$$\langle k | \tau_{\mu\nu} | k' \rangle = \langle k | \tau^0_{\mu\nu} | k' \rangle + \int \frac{dk''}{(2\pi)^3} \sum_{\lambda\sigma} \langle k | \tau^0_{\mu\lambda} | k'' \rangle$$

$$\times \Gamma_{\lambda\sigma}(k'', E) \langle k'' | \tau_{\sigma\nu} | k' \rangle . \qquad (3.56)$$

Taking into account only the first term in the expansion (3.51) we obtain the following equation for the elastic scattering amplitude of a pion on the bound state of three nucleons:

$$\langle k | \tau_{11} | k' \rangle \equiv \langle k | \tau | k' \rangle = \langle k | \tau^0 | k' \rangle$$

$$+ \varepsilon \int \frac{dk''}{(2\pi)^3} \langle k | \tau^0 | k'' \rangle \left(\frac{k''^2}{2\mu} - E \right)^{-1}$$

$$\times \left(\frac{k''^2}{2\mu} - E + \varepsilon \right)^{-1} \langle k'' | \tau | k' \rangle . \qquad (3.57)$$

Thus, instead of a complicated set of multidimensional equations for the system of four particles, a simple one-dimensional equation has been obtained, even though it applies within a limited energy range. The sole approximation (3.51) having been used can, within the considered range of energies, be perfected in a systematic way [68]. In this case it can be shown that as the absolute value of the binding energy of the ground state of the target increases, the role of the terms dropped in (3.51) becomes smaller.

We shall now demonstrate in detail how (3.57) is applied to the calculation of the $\pi^\pm$-meson scattering lengths on the ^{3}He and ^{3}H nuclei. To this end we write (3.49, 57) for the states with a total isotopic spin T of the $\pi + ^3$He system, which represents a conserved quantum number. Then, combining their solutions, we construct physical amplitudes, for example, for the ^{3}He system, the wave function of which is not an eigenfunction of the operator of the squared total isospin of the system, T. To do so, we find the projection of the potential V_π onto the states of the system with the fixed isospin T. As a result, we obtain [69]

$$V_\pi^T(r_{12}, r_3) = C_1^T \sum_{i=1}^{3} a_i^1 + C_3^T \sum_{i=1}^{3} a_i^3 , \tag{3.58}$$

where

$$C_1^T = \tfrac{1}{3} \left[1 + 2(-1)^{T+1/2} \left\{ \begin{matrix} 1 & 1 & 1 \\ 1/2 & T & 1/2 \end{matrix} \right\} \right] ;$$

$$C_3^T = \tfrac{2}{3} \left[1 - (-1)^{T+1/2} \left\{ \begin{matrix} 1 & 1 & 1 \\ 1/2 & T & 1/2 \end{matrix} \right\} \right] ;$$

$\left\{ \begin{matrix} 1 & 1 & 1 \\ 1/2 & T & 1/2 \end{matrix} \right\}$ is a $6j$ symbol; while $a^{2t}\pi N$ represent the pion-nucleon potentials acting in the state with a definite isospin of the πN system, equal to $1/2$ or to $3/2$.

For a separable πN potential acting only in the S-state the expression (3.58) in the momentum representation

$$\langle k | V_\pi^N(r_{12}, r_3) | k' \rangle = \sum_{\substack{i=1,2,3 \\ \alpha=1,2}} \Lambda^\alpha \eta_i^\alpha(k) \bar{\eta}_i^\alpha(k') , \tag{3.59}$$

where

$$\eta_i^\alpha(k) = \frac{\exp(ikz_i)}{k^2 + \beta_\alpha^2} ; \quad z_1 = \tfrac{1}{2}r_{12} + \tfrac{1}{3}r_3 ; \quad z_2 = -\tfrac{1}{2}r_{12} + \tfrac{1}{3}r_3 ;$$

$$z_3 = -\tfrac{2}{3}r_3 ; \quad \Lambda^1 = C_1^T \lambda_S^1 ; \quad \Lambda^2 = C_3^T \lambda_S^3 ; \tag{3.60}$$

$\lambda_S^{1,3}$ are the depths of the S-wave πN-potentials [70] in the states $t_{\pi N} = \tfrac{1}{2}$ and $t_{\pi N} = \tfrac{3}{2}$, respectively.

Equation (3.49) with the potential (3.59) is readily solved:

$$\langle k | \tau^0(r_{12}, r_3) | k' \rangle = \sum_{ij,\alpha\beta} \Lambda^\alpha \eta_i^\alpha(k) [A^{-1}(E, r_{12}, r_3)]_{ij}^{\alpha\beta} \bar{\eta}_i^\beta(k') . \tag{3.61}$$

The matrix elements of A are given by integrals of the product of the two form factors η_i^α and the free Green function, as it is conventionally obtained in finding the scattering amplitude on fixed centers. The most cumbersome thing in this approach is the calculation of multidimensional integrals in (3.55). Actually, this difficulty has hitherto restricted the application of this method to the heavier nuclei. In the case of three-nucleon nuclei, when the total energy is negative, the matrix A^{-1} in (3.61) represents a smooth function of its variables, i.e., of the angle between the vectors r_{12} and r_3 and of the lengths of these vectors, $|r_{12}|, |r_3|$. This simplifies significantly the determination of the matrix element of the operator τ^0.

The results of the solution of (3.57) for the (3N) scattering lengths are presented in Table 3.1.

Table 3.1.

Scattering lengths [fm]	$a^1(\pi\,^3\mathrm{He})$	$a^3(\pi\,^3\mathrm{He})$	$a(\pi^-\,^3\mathrm{He})$	$a(\pi^-\,^3\mathrm{He}\to\pi^0\,^3\mathrm{He})$
Theory	0.105	−0.210	0	−0.149
Experiment	−	−0.226	0	−

Here a^1 and a^3 represent the $\pi 3N$ scattering lengths for states with a total isospin T of the system equal to 1/2 and 3/2, respectively.

A few words must be said about what is meant in the table by "Experiment" and about the origin of the numbers in that line. Let us start with the $\pi^-\,^3\mathrm{He}$-scattering length. This quantity (just like the $\pi^+\,^3\mathrm{He}$-scattering length) is not a quantity that can be measured directly. The measurable quantity is the level shift ΔE in the ground state of the π-mesic atom $\pi^-\,^3\mathrm{He}$ which is due to the strong πN interaction. Further, utilizing Deser's formula $a_{\pi A} = \kappa\Delta E$, which is obtained by application of the perturbation theory in strong πN interaction, one finds the pion-nucleus scattering length $a_{\pi A}$. Here $\kappa = [-(2\pi(\mu_{\pi A})|\psi_0(0)^2]^{-1}$, and $\psi_0(0)$ is the wave function of the ground state of the πA atom at $r = 0$. However, in doing so one must bear in mind that the contribution to the experimental value of $\mathrm{Re}\,a_{\pi A}$ is due to two factors. The first are the purely multiple π-meson scattering processes on a nucleon of the nucleus, which can be described in the potential language and to which the theory presented above is applicable. The second are processes involving disappearance of the π-meson in intermediate states. At present there exists no rigorous theory of the latter processes. (Attempts of an approximate description of pion absorption are described below.) The Brueckner hypothesis [71] assumes that

$$\Delta E_{\mathrm{exp}} = \Delta E_{\mathrm{el}} + 1.05\,\Gamma_{\mathrm{exp}} , \qquad (3.62)$$

i.e., that the shift of the level in the π-mesic atom represents the sum of the shift due only to multiple scattering processes of the pion and the experimental value of the width of this level. Let us turn to the observable data for the $\pi^-\,^3\mathrm{He}$ atom [72]:

$$E_{\mathrm{exp}} = (0.050 \pm 0.005)m_\pi^{-1} ; \quad \Gamma_{\mathrm{exp}} = (0.034 \pm 0.012 m_\pi^{-1} .$$

Hence, utilizing (3.62) we conclude that within the experimental uncertainties $\Delta E_{\mathrm{el}} \approx 0$. Thus, the agreement between the theory and the experiment, shown in Table 3.1, must be understood in this sense.

Let us now generalize the approximate four-particle equations to the case when the target breaks up and three fragments remain in the final state. The presented method may serve as an alternative to the existing cluster models of nuclear reactions.

We shall again consider processes such as $\pi^+\,^3\mathrm{He} \to p + d + \pi$. Taking into account the three-particle nature of the final state we represent the initial Hamiltonian H in the form

$$H = H_0 + H_d + v_1 + v_2 + v_3 . \qquad (3.63)$$

Here H_0 is the sum of the kinetic energies of the relative motion of the three particles p, d, and π:

$$H_d = h_d + V_{N_1 N_2} ; \quad v_1 = V_{\pi N_3} ;$$
$$v_2 = V_{\pi N_2} + V_{\pi N_2} ; \quad v_3 = V_{N_1 N_3} + V_{N_2 N_3} ; \tag{3.64}$$

where h_d is the kinetic energy of the relative motion of the nucleons in the deuteron; $V_{\pi N_i}$ is the interaction potential of the pion and the i-th nucleon; $V_{N_i N_j}$ is the interaction potential between the i-th and j-th nucleons. As in the case of elastic scattering, we introduce several auxiliary Green functions:

$$g(Z) = (Z - H)^{-1} ; \quad \bar{g}(Z) = (H - H_d - Z)^{-1} ;$$
$$g_d(Z) = (H_0 + H_d - Z)^{-1} ; \quad g_0(Z) = (H_0 - Z)^{-1} ; \tag{3.65}$$
$$g_{00}(Z) = (H_0 + h_d - Z)^{-1} .$$

Following Faddeev, we introduce, as in Chap. 2, the operators $M_{\alpha\beta}$ and $M^0_{\alpha\beta}$:

$$M_{\alpha\beta} = v_\alpha \delta_{\alpha\beta} - v_\alpha g(Z) v_\beta ; \tag{3.66}$$

$$M^0_{\alpha\beta} = v_\alpha \delta_{\alpha\beta} - v_\alpha \bar{g}(Z) v_\beta, \alpha, \beta = 1, 2, 3 . \tag{3.67}$$

Operator $M^0_{\alpha\beta}$ represents a new object, the three-particle analogue of the t-matrix for the scattering on fixed centers τ^0 introduced above in describing elastic scattering.

One can readily verify that the operators $M^{\alpha\beta}$ and $M^0_{\alpha\beta}$ satisfy equations which formally have the form of the Faddeev equations:

$$M_{\alpha\beta} = t_\alpha \delta_{\alpha\beta} - t_\alpha g_d(Z) \sum_{\lambda \neq \alpha} M_{\lambda\beta} , \tag{3.68}$$

$$M^0_{\alpha\beta} = t^0_\alpha \delta_{\alpha\beta} - t^0_\alpha g_0(Z) \sum_{\lambda \neq \alpha} M^0_{\lambda\beta} , \tag{3.69}$$

where the "channel" t-matrices t_α and t^0_α satisfy equations of the Lippmann–Schwinger type:

$$t_\alpha = v_\alpha - v_\alpha g_d(Z) t_\alpha ; \tag{3.70}$$

$$t^0_\alpha = v_\alpha - v_\alpha g_0(Z) t^0_\alpha . \tag{3.71}$$

Obviously, when $\alpha = 2, 3$, (3.71) is an equation for a two-center t-matrix.

Bearing in mind the relations

$$T = \sum_{\alpha,\beta} M_{\alpha\beta} \quad \text{and} \quad \tilde{T} = \sum_{\alpha,\beta} M^0_{\alpha\beta} , \tag{3.72}$$

one can rewrite the Lippmann–Schwinger equation for the total four-particle transition operator T as

$$T = \tilde{T} + \tilde{T} g_0(Z) H_\mathrm{d} g_\mathrm{d}(Z) T \ . \tag{3.73}$$

This equation is a generalization of (3.48), meaning that in its explicit form it takes into account possible reaction channels through the structure of $\tilde{T}$. It is not, however, a Fredholm equation since its kernel contains disconnected blocks due to the Born term in $\tilde{T}$, which can be easily verified by iteration. Now we shall perform the Faddeev rearrangement of (3.73) to eliminate these disconnected blocks. To this end, instead of the operators $M^0_{\alpha\beta}$, we introduce the operators $W^0_{\alpha\beta}$:

$$M^0_{\alpha\beta} = t^0_\alpha \delta_{\alpha\beta} + W^0_{\alpha\beta} \ . \tag{3.74}$$

Then the operator T becomes

$$\tilde{T} = \sum_{\alpha\beta} M^0_{\alpha\beta} = \sum_\alpha t^0_\alpha + \sum_{\alpha\beta} W^0_{\alpha\beta} \equiv T^{00} + T_\mathrm{F} \ , \tag{3.75}$$

i.e. the part T^{00} is singled out explicitly, which causes the non-Fredholm character of (3.73).

As always, when rearrangement is performed, we introduce the component of the amplitudes:

$$\begin{aligned}
T_1 &= T^{00} + T^{00} g_0(Z) H_\mathrm{d} g_\mathrm{d}(Z)(T_1 + T_2) \\
T_2 &= T_\mathrm{F} + T_\mathrm{F} g_0(Z) H_\mathrm{d} g_\mathrm{d}(Z)(T_1 + T_2) \\
T_{1\alpha} &= t^0_\alpha + t^0_\alpha g_0(Z) H_\mathrm{d} g_\mathrm{d}(Z)(T_1 + T_2) \ .
\end{aligned} \tag{3.76}$$

Obviously, the following relations hold:

$$T = T_1 + T_2 \quad \text{and} \quad \sum_\alpha T_{1\alpha} = T_1 \ . \tag{3.77}$$

Substituting (3.75) into (3.73) and making use of the introduced components (3.76) we obtain a set of Fredholm equations

$$\begin{aligned}
T_{1\alpha} &= t_\alpha + t_\alpha g_0(Z) H_\mathrm{d} g_\mathrm{d}(Z) T_2 + t_\alpha g_0(Z) H_\mathrm{d} g_\mathrm{d}(Z) \sum_{\gamma \neq \alpha} T_{1\alpha} \\
T_2 &= T_\mathrm{F} + T_\mathrm{F} g_0(Z) H_\mathrm{d} g_\mathrm{d}(Z)(T_2 + \sum_\alpha T_{1\alpha}) \ .
\end{aligned} \tag{3.78}$$

From the form of the Born term for the operator

$$W^{00}_{\alpha\beta} = \begin{cases} 0 \ , & \alpha = \beta \\ t_\alpha g_0 t_\beta \ , & \alpha \neq \beta \end{cases}$$

it is clear that T_F contains no disconnected diagrams, i.e., the kernel of the second equation of the set (3.78) is also a Fredholm kernel.

Thus, a set of the four-particle equations has been obtained which takes into account the three-particle character of the final state and which is written in a form convenient for the application of approximation such as (3.51). However, before proceeding to utilize these approximations, we shall take advantage of one more simplifying factor. From the definition of the operator $W^0_{\alpha\beta}$ it follows that it may be small (for instance, with respect to the norm) if all three clusters in the final state do not approach each other simultaneously at short distances. In this case one can set

$$M^0_{\alpha\beta} \approx t^0_\alpha \delta_{\alpha\beta} \tag{3.79}$$

and (3.78) is substantially simplified:

$$T_{1\alpha} = t_\alpha + t_\alpha g_0(Z) H_d g_d(Z) \sum_{\beta \neq \alpha} T_{1\beta}, \quad T = \sum_{\alpha=1}^{3} T_{1\alpha}; \quad T_2 = 0 . \tag{3.80}$$

Thus, for the four-particle amplitudes $T_{1\alpha}$ an analogue of the Faddeev equations (3.80) has been obtained. Now we shall show that in the approximation

$$H_d \simeq \sum_{\mu\nu}^{N} |\chi_\mu\rangle \kappa_{\mu\nu} \langle \chi_\nu| \tag{3.81}$$

the equations (3.80) reduce to a set of two-dimensional integral equations. We introduce mixed Jacobi variables:

$$
\begin{aligned}
k_1 &= \frac{m p_\pi - m_\pi p_N}{m + m_\pi} , \\
k_2 &= \frac{m_\pi p_d - 2m p_\pi}{2m + m_\pi} , \quad
\begin{aligned}
q_1 &= p_d \\
q_2 &= p_N \\
q_3 &= p_\pi \\
r &= r_{12}
\end{aligned} \\
k_3 &= \frac{2m p_N - m p_d}{3m} , \\
p_d &+ p_\pi + p_N = 0 ,
\end{aligned}
\tag{3.82}
$$

and write the functions occurring in (3.80) in terms of these variables. For the Green function $g_d(Z)$ in the approximation (3.81)

$$\langle k_\alpha, q_\alpha, r | g_d(Z) | k_\alpha, q'_\alpha, r' \rangle = (2\pi)^6 \delta(k_\alpha - k'_\alpha) \delta(q_\alpha - q'_\alpha)$$
$$\times \left[\delta(r - r') \left(\frac{k^2_\alpha}{2\mu_\alpha} + \frac{q^2_\alpha}{2\bar{\mu}_\alpha} - Z \right)^{-1} - \sum_{\mu\nu} \chi_\mu(r) \Gamma^\alpha_{\mu\nu}(k_\alpha, q_\alpha) \bar{\chi}_\nu(r) \right], \tag{3.83}$$

where

$$\Gamma^\alpha_{\mu\nu}(k_\alpha, q_\alpha) = \left(\frac{k^2_\alpha}{2\mu_\alpha} + \frac{q^2_\alpha}{2\bar{\mu}_\alpha} - Z \right)^{-1} \sum_\lambda \kappa_{\mu\lambda} \left[\frac{k^2_\alpha}{2\mu_\alpha} + \frac{q^2_\alpha}{2\bar{\mu}_\alpha} - Z + \hat{I}\hat{\kappa} \right]^{-1}_{\lambda\nu} ;$$
$$\tag{3.84}$$

m is the mass of the nucleon;

$$\mu_1^{-1} = m^{-1} + m_\pi^{-1} \; ; \; \bar{\mu}_1^{-1} = (2m)^{-1} + (m + m_\pi)^{-1} \; ; \quad \mu_2^{-1} = (2m)^{-1} + m_\pi^{-1} \; ;$$

$$\bar{\mu}_2^{-1} = m^{-1} + (2m + m_\pi)^{-1} \; ; \quad \mu_3^{-1} = ((2/3)m)^{-1} \; ; \quad \bar{\mu}_3^{-1} = m_\pi^{-1} + (3m)^{-1} \; .$$

The expressions for t_α in the space of four-particle states have the form

$$\langle k_\alpha, q_\alpha, r | t_\alpha | k'_\alpha, q'_\alpha, r' \rangle = (2\pi)^2 \delta(q_\alpha - q'_\alpha) \langle k_\alpha, r | \tau_\alpha(q_\alpha) | k'_\alpha, r' \rangle \; . \quad (3.85)$$

Starting from the equation for t_α we obtain equations for the matrix elements τ_α, analogous to (3.56):

$$\langle k_\alpha, r | \tau_\alpha(q_\alpha) | k'_\alpha, r' \rangle = \delta(r - r') \langle k_\alpha | \tau_\alpha^0(r) | k'_\alpha \rangle$$
$$+ \sum_{\mu\nu} \int \frac{dk''_\alpha}{(2\pi)^3} dr'' \langle k_\alpha | \tau_\alpha^0(r) | k''_\alpha \rangle \chi_\mu(r)$$
$$\times \Gamma_{\mu\nu}^\alpha(k''_\alpha, q_\alpha) \chi_\nu(r'') \langle k_\alpha, r'' | \tau_\alpha(q_\alpha) | k'_\alpha, r' \rangle \; , \quad (3.86)$$

where $\langle k_\alpha | \tau^0(r) | k' \rangle$ is a two-center t-matrix.

Introducing the matrix elements

$$\langle k_\alpha | \tau_{\alpha,\mu\nu}^0 | k'_\alpha \rangle \equiv \int dr \, \bar{\chi}_\mu(r) \langle k_\alpha | \tau_\alpha^0(r) | k'_\alpha \rangle \chi_\nu(r) \; ;$$
$$\langle k_\alpha | \tau_{\alpha,\mu\nu} | k'_\alpha \rangle \equiv \int dr \, dr' \chi_\mu(r) \langle k_\alpha, r | \tau_\alpha(q_\alpha) | k'_\alpha, r' \rangle \chi_\nu(r') \; , \quad (3.87)$$

we obtain from (3.86) a set of one-dimensional integral equations for the matrix elements of the operators $\tau_{\alpha,\mu\nu}$:

$$\langle k_\alpha | \tau_{\alpha,\sigma\lambda}(q_\alpha) | k'_\alpha \rangle = \langle k_\alpha | \tau_{\alpha,\sigma\lambda}^0 | k'_\alpha \rangle + \sum_{\mu\nu} \int \frac{dk''_\alpha}{(2\pi)^3}$$
$$\times \langle k_\alpha | \tau_{\alpha,\sigma\mu}^0 | k''_\alpha \rangle \Gamma_{\mu\nu}^\alpha(k''_\alpha, q) \langle k''_\alpha | \tau_{\alpha,\nu\lambda}(q_\alpha) | k'_\alpha \rangle \; . \quad (3.88)$$

Thus, all the quantities entering into (3.80) have been defined, and now we may proceed to write the explicit form of the integral equations for the matrix elements of the sought operators $T_{1\alpha}$, starting from (3.80). We introduce the amplitudes $T_{\mu\nu}^\alpha$:

$$\langle k_\alpha, q_\alpha | T_{\mu\nu}^\alpha | k_\alpha, q_\alpha \rangle \equiv \int dr \, dr' \chi_\mu(r) \langle k_\alpha, q_\alpha, r | T_{1\alpha} | k'_\alpha, q'_\alpha, r' \rangle \chi_\nu(r') \; . \quad (3.89)$$

Then, (3.80) assume the form of a set of Faddeev equations for the components of the transition operators T^α with one of the three clusters (in this case, the deuteron) having internal degrees of freedom characterized by the indices μ:

$$
\begin{aligned}
\langle k_1, q_1 | T^{(1)}_{\mu\nu} | k_1', q_1' \rangle = {}& (2\pi)^3 \delta(q_1 - q_1') \langle k_1 | \tau_{1,\mu\nu} | k_1' \rangle \\
& - \sum_{\sigma\lambda} \int \frac{dk_1''}{(2\pi)^3} \langle k_1 | \tau_{1,\mu\sigma} | k_1'' \rangle \Gamma^{(1)}_{\sigma\lambda}(k_1'', q_1) \\
& \times [\langle -\alpha_2 k_1'' + \beta_2 q_1; -k_1'' - \gamma_2 q_1 | T^{(2)}_{\lambda\nu} | \\
& \quad - \alpha_2 k_1' + \beta_2 q_1; -k_1' - \gamma_2 q_1' \rangle \\
& + \langle -\alpha_3 k_1'' - \beta_3 q_1; k_1'' - \gamma_3 q_1 | T^{(3)}_{\lambda\nu} | \\
& \quad - \alpha_3 k_1' - \beta_3 q_1; k_1' - \gamma_3 q_1' \rangle] ;
\end{aligned}
$$

$$
\begin{aligned}
\langle k_2, q_2 | T^{(2)}_{\mu\nu} | k_2', q_2' \rangle = {}& (2\pi)^3 \delta(q_2 - q_2') \langle k_2 | \tau_{2,\mu\nu} | k_2' \rangle \\
& - \sum_{\sigma\lambda} \int \frac{dk_2''}{(2\pi)^3} \langle k_2 | \tau_{2,\mu\sigma} | k_2'' \rangle \Gamma^{(2)}_{\sigma\lambda}(k_2'', q_2) \\
& \times [\langle -\gamma_2 k_2'' - \beta_2 q_2; k_2'' - \alpha_2 q_2 | T^{(1)}_{\lambda\nu} | \qquad\qquad (3.90) \\
& \quad - \gamma_2 k_2' - \beta_2 q_2'; k_2' - \alpha_2 q_2' \rangle \\
& + \langle -\gamma_1 k_2'' + \beta_1 q_2; -k_2'' - \alpha_1 q_2 | T^{(3)}_{\lambda\nu} | \\
& \quad - \gamma_1 k_2' + \beta_1 q_2'; -k_2' - \alpha_1 q_2' \rangle] ;
\end{aligned}
$$

$$
\begin{aligned}
\langle k_3, q_3 | T^{(3)}_{\mu\nu} | k_3', q_3' \rangle = {}& (2\pi)^3 \delta(q_3 - q_3') \langle k_3 | \tau_{3,\mu\nu} | k_3' \rangle \\
& - \sum_{\sigma\lambda} \int \frac{dk_3''}{(2\pi)^3} \langle k_3 | \tau_{3,\mu\sigma} | k_3'' \rangle \Gamma^{(3)}_{\sigma\lambda}(k_3'', q_3) \\
& \times [\langle -\gamma_3 k_3'' + \beta_3 q_3; -k_3'' - \alpha_3 q_3 | T^{(1)}_{\lambda\nu} | \\
& \quad - \gamma_3 k_3' - \beta_3 q_3'; -k_3' - \alpha_3 q_3 \rangle \\
& + \langle -\alpha_2 k_3'' - \beta_1 q_3; k_3'' - \gamma_1 q_3 | T^{(2)}_{\lambda\nu} | \\
& \quad - \alpha_1 k_3' - \beta_1 q_3'; k_3 - \gamma_1 q_3' \rangle] .
\end{aligned}
$$

The amplitudes for the breakup of ^{3}He or for elastic scattering are found, as usual, by calculating the matrix elements of the operator $T = \sum_\alpha T^{(\alpha)}$ with respect to the corresponding asymptotic states. In writing (3.90) the following notation for the numerical constants was adopted:

$$
\alpha_1 = \frac{2m}{2m + m_\pi} ; \quad \beta_1 = \frac{2}{3}\frac{3m + m_\pi}{2m + m_\pi} ; \quad \gamma_1 = \frac{1}{3} ;
$$

$$
\alpha_2 = \frac{2m}{2m + m_\pi} ; \quad \beta_2 = \frac{m_\pi(m_\pi + 3m)}{(m + m_\pi)(2m + m_\pi)} ; \quad \gamma_2 = \frac{m}{m + m_\pi} ;
$$

$$
\alpha_3 = \frac{2}{3} ; \quad \beta_3 = \frac{3m + m_\pi}{3(m + m_\pi)} ; \quad \gamma_3 = \frac{m_\pi}{m + m_\pi} .
$$

Once more we note that the approximate four-particle equations proposed here can also be applied (and this was partly done in [73]) to describe the interaction of other hadrons, for example Λ-particles, antiprotons, K-mesons with the lightest nuclei at low energies.

3.4 Scattering of n(p) by Nuclei Consisting of Three Nucleons at Low Energies

We shall now consider a generalization of the scheme of approximate few-particle equations in the case of identical particles, i.e., we shall be interested in the description of such processes as elastic scattering of nucleons on d, ^{3}He, ^{3}H nuclei.

At present there exist practically no reasonably reliable calculations of the wave functions of such four-particle systems as, e.g., a thermal neutron + three-nucleon nucleus. In these cases the approach being presented, if generalized to identical particles, turns out to be particularly effective.

We shall now demonstrate a device allowing us to determine the identity of particles on an example with three particles. We shall consider quartet nd-scattering. We consider the nucleons to be identical by constructing the antisymmetric asymptotic functions

$$|\varphi_k\rangle = \sqrt{3}\hat{A}|\chi_d k\rangle . \tag{3.91}$$

Here $|\chi_d k\rangle$ is an unsymmetrized channel function of the n+d system; the operator

$$\hat{A} = (1/6)(1 - P_{12} - P_{13} - P_{23} + P_{12}P_{13} + P_{12}P_{23}) , \tag{3.92}$$

and P_{ij} is the operator permuting all the coordinates of particles i and j. Then for the nd-scattering amplitude we obtain

$$f(k; k', Z) = -\frac{\mu}{2\pi}\langle\varphi_{k'}|T|\varphi_k\rangle$$
$$= -\frac{\mu}{2\pi}\langle\chi_d k'|T(1 - 2P_{13})|\chi_d k\rangle .$$

Thus, in order to determine the elastic nd-scattering amplitude it is necessary to find the matrix elements over the non-antisymmetrized wave functions of the operator $\Gamma = T(1 - 2P_{13})$.

Using (3.48) for the operator T we obtain for the operator Γ

$$\Gamma = \Gamma^0 + T^0(G_0 - G_c)\Gamma \quad \Gamma^0 = T^0(1 - 2P_{13}) . \tag{3.93}$$

Turning to the approximation (3.51) and retaining, for simplicity, only the first term, we obtain the equation for the sought matrix elements of the operator Γ:

$$\langle\chi_d k'|\Gamma|\chi_d k\rangle = \langle\chi_d k'|\Gamma^0|\chi_d k\rangle$$
$$+ \varepsilon_d \int \frac{dk''}{(2\pi)^3} \frac{\langle\chi_d k'|T^0|\chi_d k''\rangle}{(E'' - Z)(E'' - Z + \varepsilon_d)}\langle\chi_d k''|\Gamma|\chi_d k\rangle. \tag{3.94}$$

If for calculating T^0 one uses the triplet NN-potential leading to the low-energy NN characteristics $\varepsilon_d = -2.225\,\text{MeV}$, $a_t = 5.378\,\text{fm}$, and $r_t = 2.7\,\text{fm}$, then the solution to (3.94) will lead to the quartet nd-scattering length $^4a = 5.25\,\text{fm}$. The experimental value is $6.35 \pm 0.02\,\text{fm}$. Since the nearest ^{3}He breakup threshold energy is nearly 3 times as large as the binding energy of the deuteron, one may expect application to the n^3He system of the one-term approximation (3.51) to lead to scattering lengths that will be closer to the experimental values.

Note that, since in n^3He scattering the channel $n^3He \rightarrow p^3H$ is always open (owing to $|\varepsilon_t| > |\varepsilon_{3_{He}}|$), the expressions for elastic n^3He-scattering lengths have an imaginary part. Usually, in solving Yakubovsky equations for this process one neglects the Coulomb interaction between the protons because of computational difficulties. As a consequence, the imaginary parts of the scattering lengths cannot be calculated. Utilization of the approximate four-particle equations makes it possible to partly overcome this drawback.

By reasoning just like in the three-particle case, one can obtain the following expression for the elastic n^3He-scattering amplitude:

$$f(\mathbf{k}', \mathbf{k}, Z) = -(\mu/2\pi)\langle \chi_{^3He}\mathbf{k}'|T(1 - 3P_{34})|\chi_{^3He}\mathbf{k}\rangle ,$$
$$\mu = (3/4)m . \tag{3.95}$$

Thus, once more the matrix elements of the new operator

$$\Gamma \equiv T(1 - 3P_{34}) . \tag{3.96}$$

are needed, instead of the transition operator T. As in the three-particle case, we obtain for the operator Γ

$$\Gamma = \Gamma^0 + T^0(G_0 - G_c)\Gamma , \quad \Gamma^0 = T^0(1 - 3P_{34}) . \tag{3.97}$$

Since there exist only two nuclei consisting of three nucleons, it is natural to approximate the Hamiltonian of the three-nucleon subsystem of the total Hamiltonian H by the two-term formula (3.51)

$$h_c \approx h_c^{(2)} = \varepsilon_{^3He}|\chi_{^3He}\rangle\langle\chi_{^3He}| + \varepsilon_{^3He}|\chi_{^3H}\rangle\langle\chi_{^3H}| . \tag{3.98}$$

For simplicity, and without losing much accuracy, we shall assume the spatial parts of the wave functions of the nuclei ^3He and ^3H to be identical:

$$\langle \mathbf{123}|\chi_{^3He}\rangle \equiv \chi_{^3He}(\mathbf{r}_{12}, \mathbf{r}_3) = N\exp(-\kappa\varrho)\xi_{1/2} ;$$
$$\langle \mathbf{123}|\chi_{^3H}\rangle \equiv \chi_{^3H}(\mathbf{r}_{12}, \mathbf{r}_3) = N\exp(-\kappa\varrho)\xi_{-1/2} , \tag{3.99}$$

where

$$\kappa = \kappa_{^3He} ; \quad \varrho^2 = (3/2)r_{12}^2 + 2r_3^2 ;$$

$\xi_{\pm 1/2}$ is the antisymmetrized spin-isospin function of three nucleons in the state with spin $1/2$ and isospin projection $\pm 1/2$.

Let S and T be the total spin and isospin, respectively, of the system of four nucleons. Then the scattering amplitude in the channel with spin S is given by

$$f_{n^3He}^S(\mathbf{k}', \mathbf{k}, Z) = \frac{\mu}{2\pi}\tfrac{1}{2}[\langle\chi\mathbf{k}'|\Gamma_{ST=0}(Z)|\chi\mathbf{k}\rangle + \langle\chi\mathbf{k}'|\Gamma_{ST=1}(Z)|\chi\mathbf{k}\rangle] ,$$
$$S = 0, 1 ; \quad \langle\mathbf{r}_{12}, \mathbf{r}_3|\chi\rangle = N\exp(-\kappa\varrho) . \tag{3.100}$$

The matrix elements of the operator Γ in the approximation (3.98) are

$$\langle \chi k' | \Gamma_{ST} | \chi k \rangle = \langle \chi k' | \Gamma_{ST}^0 | \chi k \rangle + \int \frac{dk''}{(2\pi)^3} \frac{R(k'', Z)}{E'' - Z}$$
$$\times \langle \chi k' | T_{ST}^0 | \chi k'' \rangle \langle \chi k'' | \Gamma_{ST} | \chi k \rangle , \qquad (3.101)$$

where

$$R(k'', Z) = \frac{1}{2} \frac{\varepsilon_{3He}}{E'' - Z + \varepsilon_{3He}} + \frac{1}{2} \frac{\varepsilon_{3He}}{E'' - Z + \varepsilon_{3He}} . \qquad (3.102)$$

An important feature of the structure of $R(k'', Z)$ is the presence of a pole at $Z = \varepsilon_{3He}$, leading to the appearance of the imaginary terms in $R(k'', Z)$ and, correspondingly, in the $n\,^3$He-scattering lengths. In $p\,^3$He-scattering the corresponding scattering lengths are real since in this case one must set $Z = \varepsilon_{3H}$ in (3.101, 102). Thus the function $R(k'', Z)$ has no imaginary part.

Table 3.2.

ST	$h_c \approx h_c^{(2)}$	Yakubovsky equations	ST	$h_c \approx h_c^{(3)}$	[61]
00	8.1 − 1.5i	12.34	01	4.0 + 0.06i	3.77
10	2.9 + 0.1i	3.03	11	5.6 − 0.09i	3.13

Table 3.3.

A	$h_c \approx h_c^{(2)}$	[62]	Experiment [74]
A_0	6.05 − 0.72i	8.05	$6.1 \pm 0.6 - (4.4448 \pm 9 \times 10^{-4})$i
A_1	4.25 + 0.005i	3.08	$4.0 \pm 0.2 - (1.7 \pm 0.8 \times 10^{-8})$i

Note: the $n\,^3$He scattering lengths $A_S(n\,^3He) = \frac{1}{2}(A_{S0} + A_{S1})$.

Table 3.4.

A	$h_c \approx h_c^{(2)}$	[62]	Experiment [75]
A_0	3.8	3.77	3.31 ± 0.12
A_1	4.9	3.13	3.60 ± 0.10

Note: the $n\,^3$H scattering lengths $A_S(n\,^3H) = A_{S1}$.

In calculating the kernel and the inhomogeneous term in (3.101) we shall use the triplet NN-potential, that has been already applied in nd scattering, as well as the singlet potential with parameters reproducing the singlet NN scattering length $a_s = -23.69$ fm.

The results obtained from solution of (3.101) for scattering lengths in various spin states are presented in Tables 3.2–4. For comparison, the results based on solution of the Yakubovsky equations [62] and the experimental scattering lengths (fm) are also given.

As can be seen from Tables 3.3, 4, in the case of four nucleons, the approximate four-particle equations (3.101) describe the interaction of neutrons with nuclei consisting of three nucleons, at low energies quite well.

3.5 NN-Scattering in the Nonrelativistic Quark Model (The Six-Body Problem)

Finite-dimensional approximations of the Hamiltonians of subsystems make it possible to deal with significantly more complicated problems than the ones considered above. In this section we discuss this approach for collisions between composite nucleons, i.e., we proceed with the solution of the six-body problem [76].

The key condition for the problem to be tractable is that no quarks leave the nucleon. This allows the approximation of intranucleon Hamiltonians by operators of finite rank and permits one to write the dynamic equations with respect to a single variable, the momentum of the relative motion of the centers of mass of the colliding objects.

Let H_0 be the free Hamiltonian of the relative motion of the nucleon centers of mass; $H_i(i = 1, 2)$ is the Hamiltonian of internal motion of the quarks in the ith nucleon; $V = \sum_{i=1}^{3} \sum_{j=4}^{6} V_{ij}$, and V_{ij} is the interaction potential between quarks belonging to different nucleons. We also introduce Hamiltonians of the subsystems

$$H_c \equiv H_1 + H_2, H_{\mathrm{acc}} = H_0 + H_c . \tag{3.103}$$

We follow the procedure previously applied in obtaining the approximate four-particle equations for elastic scattering. To this end we introduce the Green functions

$$\begin{aligned} G_0(Z) &= (H_0 - Z)^{-1}; & G_{\mathrm{acc}}(Z) &= H_{\mathrm{acc}} - Z)^{-1}; \\ \tilde{G}(Z) &= (H_0 + V - Z)^{-1}; & G(Z) &= (H - Z)^{-1}. \end{aligned} \tag{3.104}$$

where $H = H_0 + H_c + V$ is the total Hamiltonian, and the auxiliary operator T^0 is given by

$$T^0 = V - V\tilde{G}(Z)V = V - VG_0(Z)T^0 . \tag{3.105}$$

This model is extremely simplified because of neglect of atomic degrees of freedom.

The operator T^0 defined by (3.105) is the generalization of the four-particle T^0. By definition, T^0 describes the scattering of three fixed centers (quarks having nucleon number 1, for instance) on three other fixed centers. This problem can also be solved analytically if the quark-quark potential can be approximated by operators of finite rank.

Using the adopted notation we obtain, as usual, for the total operator T

$$T = T^0 + T^0 G_0(Z) H_c G_{acc}(Z) T = T^0 + T^0 [G_0(Z) - G_{acc}(Z)] T . \qquad (3.106)$$

We approximate the Hamiltonian of the subsysten H_c by a first-rank operator

$$H_c \approx H_c^{(1)} = \varepsilon_0 |\chi_1 \chi_2\rangle \langle \chi_1 \chi_2| , \qquad (3.107)$$

where $|\chi_1 \chi_2\rangle = |\chi_1\rangle \chi_2\rangle$, $|\chi_i\rangle$ is the eigenfunction of the internal Hamiltonian H_i; $\varepsilon_0/2$ is the eigenvalue of H_i corresponding to the ground state of the nucleon; $\varepsilon_0 = 2(m_N - 3m_q)$; m_N is the mass of a nucleon; $m_q = 340\,\text{MeV}$ is the mass of a quark. Approximation (3.107) can readily be made better by adding a term corresponding to the excitation of the isobar. Such a procedure undoubtedly extends the range of energies at which one may apply the approximate few-particle equations. As in the four-particle case, we obtain a one-dimensional integral equation for the elastic NN-scattering amplitude:

$$\langle k|\tau|k'\rangle = \langle k|\tau^0|k'\rangle$$
$$+ \varepsilon_0 \int \frac{dq}{(2\pi)^3} \langle k|\tau^0|q\rangle G_0(q, Z) G_0(q, Z - \varepsilon_0)\langle q|\tau|k'\rangle . \qquad (3.108)$$

In this case, however, the matrix element $\langle k|\tau^0|k'\rangle$ is a significantly more complicated integral (compared to the case of four particles):

$$\langle k|\tau^0|k'\rangle = \int \chi_1^*(x_1)\chi_2^*(x_2)\langle k|\tau^0(x_1, x_2)|k'\rangle \chi_1(x_1)\chi_2(x_2) dx_1 dx_2 . \qquad (3.109)$$

In (3.108) the following notation is utilized:

$$\langle k|\tau|k'\rangle \equiv \int \chi_1^*(x_1)\chi_2^*(x_2)\langle k, x_1, x_2, |T|k', x_1', x_2'\rangle$$
$$\times \chi_1(x_1')\chi_2(x_2') dx_1 dx_2 dx_1' dx_2' ; \qquad (3.110)$$

$$\langle k, x_1, x_2|T^0|x_1', x_2', k'\rangle = \delta(x_1 - x_1')\delta(x_2 - x_2')\langle k|\tau^0(x_1, x_2)|k'\rangle . \qquad (3.111)$$

$x_1 \equiv \{r_{23}, r_{123}\}$, $x_2 \equiv \{r_{56}, r_{456}\}$; r are the Jacobi variables (Fig. 3.9).

Using (3.111) we obtain the equation for the auxiliary amplitude $\langle k|\tau^0(x_1, x_2)|k'\rangle$:

$$\langle k|\tau^0(x_1, x_2)|k'\rangle = \langle k|V|k'\rangle \sum_{m=1}^{9} \varphi_m(x_1, x_2, k)\varphi_m^*(x_1, x_2, k')$$
$$- \sum_{m=1}^{9} \varphi_m(x_1, x_2, k) \int \frac{dq}{(2\pi)^3} \varphi_m(x_1, x_2, q)$$
$$\times \langle k|V|q\rangle G_0(q_1 Z)\langle q|\tau^0(x_1, x_2)|k'\rangle , \qquad (3.112)$$

where the functions φ_m have the form

$$\varphi_m(x_1, x_2, k) = \prod_{\substack{l=23,123 \\ 56,456}} \exp[i\alpha_m^l r_l k] , \qquad (3.113)$$

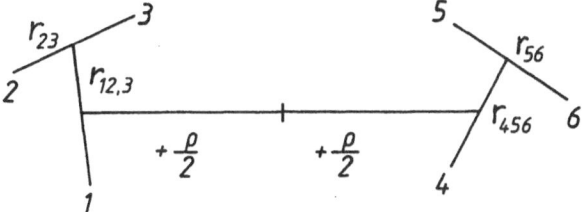

Fig. 3.9. Jacobi variables for six particles

and α_m^l is a numerical matrix.

In the case of a separable potential V the solution of (3.112) can be written

$$\langle k|\tau^0(x_1, x_2)|k'\rangle = \bar{\eta}_1(x_1, x_2, k)X^{-1}(x_1, x_2, Z)\eta(x_1, x_2, k') , \qquad (3.114)$$

where η is a multidimensional vector whose components are

$$\eta_i(x_2, x_2, k) = \begin{cases} g(k')\varphi_i^*(x_1, x_2, k'), & i = 1, \dots, 9 ; \\ \tilde{g}(k')\varphi_{i-9}^*(x_1, x_2, k'), & i = 10, \dots, 18 ; \end{cases}$$

$$\eta_{1i}(x_1, x_2, k) = \begin{cases} \lambda g(k)\varphi_i(x_1, x_2, k), & i = 1, \dots, 9 ; \\ \tilde{\lambda}\tilde{g}(k)\varphi_{i-9}(x_1, x_2, k), & i = 10, \dots, 18 ; \end{cases}$$

λ and $g(k)$ are the intensity and form factor of the separable quark-quark potential. The 12-fold integral (3.109) now becomes

$$\int dx_1 dx_2 \bar{\eta}_1(x_1, x_2, k)X^{-1}(x_1, x_2, Z)\eta(x_1, x_2, k')|\chi_1(x_1)|^2|\chi_2(x_2)|^2 .$$

$$(3.115)$$

For calculating the integral in (3.115) we shall take advantage of the absence of narrow resonances in the elastic NN-scattering at energies below 1 GeV. This permits one to assume that the inverse matrix X^{-1} is a sufficiently smooth function of the variables x_1 and x_2.

Indeed, in the region of a resonance the amplitude must exhibit the property that $\langle k|\tau^0|k'\rangle \approx h(k)t(E)h(k')$. Hence, from (3.115) we arrive at the conclusion that in this case (i.e., resonance) a very sharp dependence of X^{-1} on x_1 and x_2 must exist. Roughly speaking, one should expect the relationship

$$X^{-1}(x_1, x_2, Z) \sim A(\bar{x}_1, \bar{x}_2, Z)\delta(x_1 - \bar{x}_1)\delta(x_2 - \bar{x}_2) .$$

Thus, assuming on physical grounds, the matrix X^{-1} to exhibit smooth behavior, one can obtain an approximate value for the amplitude (3.115) by expanding the matrix X^{-1} into a series in the vicinity of certain equilibrium values of $\bar{x}_1$ and $\bar{x}_2$. Below we shall take only the first term of the expansion into account, upon which integration in (3.115) is readily performed. In the range of energies dealt with (below 400 MeV), this procedure gives rise to an error that does not exceed 15%.

The subsequent solution of this problem no longer presents difficulties. Indeed, separating in (3.108) the angular variables, we arrive at the one-dimensional equation

$$T_l(k, k', Z) = T_l^0(k, k', Z) + \int_0^\infty dq \xi_l(k, q, Z) T_l(q, k', Z) , \qquad (3.116)$$

where

$$\xi_l(k, q, Z) = \frac{1}{2\pi^2} \frac{T_l^0(k, q, Z)}{[q^2/(2\mu) - Z][q^2/(2\mu) - Z + \varepsilon_0]} ;$$

$$T_l^0(k, k', Z) = (1/2) \int_{-1}^{1} \langle k | \tau^0 | k' \rangle P_l(x) dx ; \quad x = (kk')/(kk') .$$

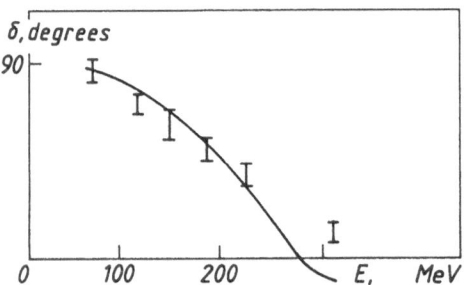

Fig. 3.10. NN scattering phases: the *solid curve* is the computed NN scattering S phase based on (3.116). The bars are the experimental values of the NN scattering $^3S_1$ phase

Equation (3.116) has been solved for $l = 0$ by applying the Bateman method with three cuts. The resulting energy dependence of the S phase is shown in Fig. 3.10. The points correspond to the experimental $^3S_1$ phase in NN scattering.

As can be seen from Fig. 3.10, (3.108) gives a qualitative idea of the dynamics of the motion of a six-quark system.

4. Related Problems

In this chapter, certain applications of the theory of few-body systems, as well as a generalization of the theory to systems with a variable number of particles are considered. Sum rules are established which represent a generalization of the known Levinson theorem and the application of these rules in statistical mechanics is discussed.

4.1 The Faddeev Equations and the Heisenberg Ferromagnetic

Let us now see how the Faddeev equations arise in the description of a Heisenberg ferromagnetic [77]. We introduce the Heisenberg Hamiltonian H_S, describing the interaction of adjacent spins (we shall be dealing with spin $1/2$) in an isotropic cubic lattice of dimension $d(d = 1, 2, 3)$:

$$H_S = -\frac{1}{2} J \sum_i \sum_\Delta S_i S_i + \Delta, J > 0 \,. \tag{4.1}$$

Here S_i is the spin operator in the i-th lattice point; summation over i indicates summation over all the lattice points, while the summation over Δ involves the nearest neighbors of the given spin situated at the i-th lattice site. (The symbol $S_i + \Delta$ stands for the spin operator of a particle situated at one of the lattice points closest to the i-th one and determined by the vector Δ.)

We shall call a state with a single flipped spin a one-magnon state. If N is the number of spins, then the operator

$$n = -\frac{1}{2} N + \sum_i S_i^z \tag{4.2}$$

defines the number of magnons in the given state and represents a good quantum number. By definition, $n = 0$ in the ground state 0. In order to derive the Faddeev equations, matrix elements are needed of the Hamiltonian H_S in the subspace of three-magnon ($n = 3$) states. For $S = 1/2$ such states have

$$|m_1, m_2, m_3\rangle \equiv S_{m_1}^+ S_{m_2}^+ S_{m_3}^+ |0\rangle \quad \text{if} \quad m_1 \neq m_2 \neq m_3 \neq m_1$$
$$|m_1, m_2, m_3\rangle \equiv 0, \quad \text{if} \quad m_i = m_k (i \neq k) \,, \tag{4.3}$$

where m_i is the vector determining the position of the i-th magnon.

The restrictions imposed on the spin projections in the definition of the three-magnon state (4.3) are inconvenient for practical calculations, therefore we shall utilize the extended space of states $|m_1, m_2, m_3\rangle$ exhibiting the following properties:

$$\langle m_1 m_2 m_3 | m_1' m_2' m_3' \rangle = \sum_{P(p_i)} \delta m_{p_1} m_1' \delta m_{p_2} m_2' \delta m_{p_3} m_3' \; ; \tag{4.4}$$

$$\frac{1}{6} \sum_{m_1, m_2, m_3} |m_1 m_2 m_3\rangle\langle m_1 m_2 m_3| = 1 \; , \tag{4.5}$$

where $\sum_{P(p_i)}$ stands for summation over all permutations 1, 2, 3. Such an extension of the space does not affect the dynamic properties of the considered spin system, since the matrix elements of H_S over the states (4.4, 5) become zero when $m_i = m_k (i \neq k)$. Indeed, taking into account the commutation relations for spin operators we find

$$H_S |m_1 m_2 m_3\rangle = J(1 - \delta_{m_1 m_2} - \delta_{m_1 m_3} - \delta_{m_2 m_3} + 2\delta_{m_1 m_2} \delta_{m_2 m_3})$$

$$\times [3d|m_1 m_2 m_3\rangle - \sum_{\Delta}(\delta_{m_1, m_2 + \Delta} + \delta_{m_1, m_1 + \Delta} + \delta_{m_2 m_2 + \Delta})$$

$$\times |m_1 m_2 m_3\rangle - \frac{1}{2} \sum_{\Delta} \sum_{P(p_i)} (1 - \delta_{m_{p_1} + \Delta, m_{p_2}} - \delta_{m_{p_1} + \Delta, m_{p_3}})$$

$$\times |m_{p_1} + \Delta, m_{p_2}, m_{p_3}\rangle] \; . \tag{4.6}$$

From (4.6) it follows that the matrix elements of H_S over functions belonging to the extension are equal to zero.

We introduce the Fourier transform of the three-magnon state:

$$|k_1, k_2, k_3\rangle = \frac{1}{N^{3/2}} \sum_{m_1 m_2 m_3} \exp\left[i \sum_n k_n m_n\right] m_1 m_2 m_3\rangle \; , \tag{4.7}$$

where k_i is the momentum of the i-th magnon. For convenience we shall denote by k the set of momenta of three-magnon state, k_1, k_2, k_3. The states $|k\rangle$ form a complete set of orthonormalized states in the subspace of three-magnon states, i.e.,

$$\langle k|k'\rangle = \sum_{P(p_i)} \delta(k_{p1} - k_1')\delta(k_{p2} - k_2')\delta(k_{p3} - k_3') \; ; \tag{4.8}$$

$$\frac{1}{6} \sum_{k_1 k_2 k_3} |k_1 k_2 k_3\rangle\langle k_1 k_2 k_3| = 1 \; . \tag{4.9}$$

Using (4.4–9) we obtain for the matrix element of H_S

$$\langle k|H_S|k'\rangle = \frac{1}{N^3} \sum_{m,m'} \exp\left[i\sum_n k'_n m'_n - i\sum_n k_n m_n\right]$$

$$\langle m_1 m_2 m_3|H_S|m'_1 m'_2 m'_3\rangle$$

$$= \sum_{n=1}^{d}[3J - J(\cos k_{1n} + \cos k_{2n} + \cos k_{3n})]\langle k|k'\rangle$$

$$+ \frac{4J}{N}\delta(P - P')\sum_{i,j=1}^{3}\sum_{n=1}^{d}\delta(k'_j - k_j)$$

$$\times \left[\cos p_{jn}\left(\cos\frac{K'_{in}}{2} - \cos p'_{in}\right) - \tfrac{1}{2}\varepsilon(p_n, K'_{in}, p'_{in})\right]$$

$$+ \delta(P - P')\sum_{n=1}^{d}\sum_{i=1}^{3}g_n^{(1)}(k)f_n^{(i)}(k') \ . \tag{4.10}$$

In (4.10) the following notation is adopted:

$$P = k_1 + k_2 + k_3 \ , \quad K_i = k_j + k_k \ , \quad p_i = \tfrac{1}{2}(k_j - k_k) \ , \tag{4.11}$$
$$i \neq j \neq k \ ; \quad i,j,k = 1,2,3 \ ;$$

$$\varepsilon(P_n, K'_{in}, p'_{in}) = 3 - \cos(P_n - K'_{in}) - 2\cos\left(\frac{K'_{in}}{2}\right)\cos p'_{in} \ ; \tag{4.12}$$

$$f_n^{(1)}(k') = \frac{4J}{N}\sum_{i=1}^{3}[2\cos(P_n - K'_{in}) - \cos K'_{in} - 1] \ ; \tag{4.13}$$

$$f_n^{(2)}(k') = \frac{4J}{N^2}\sum_{i=1}^{3}[2\sin(P_n - K'_{in}) - \sin K'_{in}] \ ; \tag{4.14}$$

$$f_n^{(3)}(k') = \frac{12J}{N^2}\sum_{i=1}^{3}[1 - \cos(P_n - K'_{in})] \ ; \tag{4.15}$$

$$g_n^{(1)}(k) = \sum_{i=1}^{3}\cos(P_n - K_{in}) \ ; \tag{4.16}$$

$$g_n^{(2)}(k)\sum_{i=1}^{3} = \sin(P_n - K_{in}) \ ; \tag{4.17}$$

$$g_n^{(3)}(k) = 1 \ . \tag{4.18}$$

The index i denotes the number of the particular magnon, while n is the number of the coordinate axis; $n = 1, \ldots, d$.

The Hamiltonian (4.10) can formally be represented

$$H_S = H_0 + \sum_{i=1}^{3}V_i + V^{(3)} \ . \tag{4.19}$$

The summands in (4.19) have the following meaning. The Hamiltonian H_0 describes the free motion of three magnons of momenta k_i and energies E_i:

$$E_i = J \sum_{n=1}^{d} (1 - \cos k_{in}) . \tag{4.20}$$

The second term is responsible for pair interactions between magnons (the magnon with number i is a spectator). Finally, the last term in (4.19) corresponds to the three-particle interaction of magnons. We note that both the pair and the three-particle interactions are separable. This property of the interaction potentials, as always, leads to it being possible to decrease the dimensionality of the Faddeev equations.

Having at our disposal the three-magnon Hamiltonian in the form of (4.19) we can now proceed to obtain the Faddeev equations for the T-matrix. Following Chap. 2, we write the three-magnon Lippmann–Schwinger equation for the transition operator T:

$$T(Z) = \sum_{i=1}^{4} V_i + \sum_{i=1}^{4} V_i G_0(Z) T(Z) , \tag{4.21}$$

where $G_0(Z) = (Z - H_0)^{-1}$, and V_4 represents the three-magnon interaction. We now define the Faddeev components of the T-matrix:

$$T^{(i)} = V_i + V_i G_0 T , \quad i = 1, 2, 3 ; \tag{4.22}$$

$$T^{(4)} = V_4 + V_4 G_0 T , \quad T = \sum_{i=1}^{4} T^{(j)} , \tag{4.23}$$

whence we directly obtain the set of the Faddeev equations for these components,

$$T^{(i)} = T_i + T_i G_0 \left(T^{(j)} + T^{(k)} + T^{(4)} \right) , \quad 1, 2, 3 = i, j, k ; \tag{4.24}$$

$$T^{(4)} = T_4 + T_4 G_0 \left(T^{(1)} + T^{(2)}z + T^{(3)} \right) ; \tag{4.25}$$

The T-matrices T_i satisfy the Lippmann–Schwinger equations

$$T_i = V_i + V_i G_0 T_i , \quad i = 1, \ldots, 4 . \tag{4.26}$$

Clearly, the separable form of the "potentials" V_i makes it possible to find the solutions of (4.26) in an analytical form. We first obtain solutions for the pair, i.e., two-magnon, T-matrices T_i ($i = 1, 2, 3$). We introduce the complete set of orthonormalized three-particle states $|PK_j p_j\rangle$:

$$\langle PK_j p_j | P'K'p_j \rangle = \delta(P - P')\delta(K_j - K'_j)[\delta(p_j - p'_j) + \delta(p_j + p'_j)] ; \tag{4.27}$$

$$\frac{1}{2} \sum_{PK_j p_j} |PK_j p_j\rangle \langle PK_j p_j| = 1 \tag{4.28}$$

and then solve (4.26) for the pair T-matrix in the basis (4.27, 28). Taking into account the separable character of (4.10) we immediately obtain

$$\langle PK_j p_j | T_j | k_j' \rangle = \delta(P - P') \sum_{i=1}^{2d} h_i(p_j) \psi_i(K_j, k', E) . \tag{4.29}$$

The vector h_i has the components

$$h(p_1) = \cos p_{11}, 1, \ldots, \cos p_{1d}, 1) , \tag{4.30}$$

while the functions ψ_i represent the solution of the corresponding set of algebraic equations; they can be written [77]

$$\psi_i = \sum_{m=1}^{2d} A_{im}(K, E) v_m(K, k') , \quad i = 1, \ldots, 2d . \tag{4.31}$$

By using the separable character of the three-particle potential (4.10) we similarly obtain for the matrix elements of the operator T_4

$$\langle k | T_4 | k' \rangle = \delta(P - P') \sum_{i=1}^{3d} g_i(k) \tau_i(k', Z) , \tag{4.32}$$

where, as usual, the $3d$-dimensional vector $g(k)$ is made up of the form factors of the separable potential V_4, i.e.,

$$g = \left(g_1^{(1)}, g_1^{(2)}, g_1^{(3)}, \ldots, g_d^{(1)}, g_d^{(2)}, g_d^{(3)} \right) , \tag{4.33}$$

and the functions τ_i satisfying a set of linear equations are

$$\tau_i = \sum_{j=1}^{3d} \alpha_{ij}(P, Z) f_j(k') , \quad i = 2, \ldots, 3d . \tag{4.34}$$

Substituting the expressions of the T-matrices T_i from (4.29, 32) into the Faddeev equations (4.24, 25) for the Faddeev components $T^{(i)}$ we obtain

$$\langle PK_i p_i | T^{(i)} | k' \rangle = \sum_{q=1}^{2d} h_q(p_i) \varphi_q(K_i, P, k', Z) , \quad i = 1, 2, 3 , \tag{4.35}$$

where functions φ_q satisfy the set of one-dimensional integral equations

$$\varphi_q(K, P, k', Z) = \varphi_q^0(K, k', Z) + \frac{1}{(2\pi)^d} \sum_{q'=1}^{2d} \int_{-\pi}^{\pi} dK'' R_{qq'} \tag{4.36}$$

$$\times (K'', K, P, Z) \varphi_{q'}(K'', P, k', Z) , \quad q = 1 \ldots 2d ,$$

where

$$\int_{-\pi}^{\pi} dK \equiv \prod_{n=1}^{d} \int_{-\pi}^{\pi} dK_n .$$

The expressions for the kernel and for the inhomogeneous term can be written

$$\varphi_q^0(K, k', Z) = \psi_q(K, k', Z) + \sum_{r=1}^{3d} \Gamma_{qr}(K, P, Z) \tau_r(k') . \tag{A}$$

Here, ψ_q is determined by (4.31), where

$$v_n(K, k') = \frac{4J}{N} \sum_{i=1}^{3} \delta_{K'_i K} \left(\cos \frac{1}{2} K_{in} - \cos P'_{in} \right) \tag{B}$$

and

$$\Gamma_{qr}(K, P, Z) = \frac{1}{6} \sum_{k} \delta_{PP'} \frac{\psi_q(K, k, Z) g_r(k)}{Z - Z(k)} ; \tag{C}$$

$$R_{qq'}(K'', K, P, Z) = N \sum_{P} \frac{\psi_q^{(1)}(K, K'') h_{q'}(p)}{Z - Z(P, K, p)}$$

$$+ \frac{3}{2} \sum_{rn} \Gamma_{qr}(K'', P, Z) \alpha_q(P, Z) l_{nq'}(K'', P, Z) . \tag{D}$$

Function $\psi_q^{(1)}$ may also be defined by (4.31) involving the change $v_n \to v_n^{(1)}$, where $v_n^{(1)}$ is the first term in the right-hand side of (B).

Similarly,

$$\langle k | T^{(4)} | k' \rangle = \delta(P - P') \sum_{m=1}^{3d} g_m(k) T_m(k', Z) . \tag{4.37}$$

Functions $T_m(k', Z)$ are expressed through the solutions of (4.36):

$$T_m(k', P, Z) = \tau_m(k') + \frac{3}{2} \int_{-\pi}^{\pi} \frac{dK}{(2\pi)^d} \sum_{n=1}^{3d} \sum_{q=1}^{2d} \alpha_{mn}(P, Z)$$

$$\times I_{nq}(K, P, Z) \varphi_q(K, P, k', Z) . \tag{4.38}$$

The expression for $I_{nq'}$ is

$$I_{nq'}(K'', P, Z) = N \sum_{p} \frac{f_n(K, p) h_{q'}(p)}{Z - Z(P, K, p)} .$$

Thus, the fourth Faddeev component of the T-matrix $T^{(4)}$ is actually expressed in the form of quadratures through the solutions generated by the pair interaction of magnons.

Equations (4.36) can only be solved by numerical integration, therefore below we shall consider the case of a one-dimensional ferromagnetic for which the three-magnon T-matrix will be found explicitly. For simplicity, we shall find the T-matrix in the vicinity of the pole in the energy corresponding to the bound state of three magnons, i.e., at E equal to the value obtained by *Bethe* [79]:

$$E = E_{\mathrm{B}} = \tfrac{1}{3}(1 - \cos P) . \tag{4.39}$$

Thus, in the vicinity of the pole the representation for the T-matrix will, obviously, be

$$T(E) \approx \frac{V(|\psi_{\mathrm{B}}\rangle\langle\psi_{\mathrm{B}}|V}{E - E_{\mathrm{B}}} , \tag{4.40}$$

where $|\psi_{\mathrm{B}}\rangle$ is the wave function of the bound state; hence, up to a normalization factor the T-matrix is

$$\langle k|T(E)|k'\rangle = \langle k|V|\psi_{\mathrm{B}}\rangle . \tag{4.41}$$

We use Bethe's wave function of the bound state $|\psi_{\mathrm{B}}\rangle$:

$$|\psi_{\mathrm{B}}\rangle = \sum_{m_1 m_2 m_3} a_{m_1 m_2 m_3} |m_1 m_2 m_3\rangle , \tag{4.42}$$

where in the limit of large N the coefficients $a_{m_1 m_2 m_3}$ are

$$a_{m_1 m_2 m_3} = \exp(vN)\exp[i(P - 2u)m_2 + iu(m_1 + m_3) - v(m_3 - m_1)] , \tag{4.43}$$

and $k'_1 = u + iv$; $k'_2 = P - 2u$; $k'_3 = u - iv$, the parameters u and v being determined by the total momentum P:

$$u = \arctan \frac{2Z}{z^2 + 3} ; \quad \exp(-2v) = \frac{1 + z^2}{9 + z^2} ; \quad z = 3\cot\frac{P}{2} . \tag{4.44}$$

Using (4.43), in the limit that $N \to \infty$ we find for the T-matrix (4.41)

$$\langle k|T|k'\rangle = \delta(P - P')$$
$$\times \sum_{i=1}^{3}\left[\frac{\cos p_i \cdot \exp(-v)}{\cosh v - \cos(u - P + K_i)}\right.$$
$$\times \left\{2[e^{-v}\cos(u + K_{i/2}) - \cos(P - \tfrac{3}{2}K_i)]\right.$$
$$+ [\cos(P - K_i) + \cos(P - 2K_i)$$
$$\left.\left.-e^{-v}\cos u - e^{-v}\cos(u - K_i)]\right\}\right] \tag{4.45}$$

Expression (4.45), as was to be expected, has only a single pole at $E = E_{\mathrm{B}}$. We note that its structure is similar to the solutions of the one-dimensional Faddeev equations for particles interacting through a δ-like potential [80].

4.2 Sum Rules and Virial Coefficients

We shall now deal with sum rules, or trace formulas, as they are called by mathematicians. In nuclear physics sum rules have been known for quite a long time in the theory of photonuclear reactions. They relate various moments of the total absorption cross section of the photon to other observables. For example, in [81] the following sum rule has been obtained on the basis of certain analytic properties of the amplitude of the process:

$$\int_0^\mu \sigma_\gamma(E)dE = 2\pi^2 \frac{e^2\hbar}{MC} \frac{ZN}{A} \left(1+0.1\frac{A^2}{NZ}\right) , \tag{4.46}$$

where σ_γ is the total absorption cross section for photons of all multipolarities by a nucleus consisting of A nucleons, Z protons, and N neutrons; μ is the threshold for pion photoproduction.

We shall now consider the sum rules that relate the eigenvalues with the integrals of scattering amplitudes, phases or interaction potentials. Practically all results available in this area have been already obtained in [82–86], which we shall follow below.

We shall start with the simplest case of the one-dimensional problem of two bodies on the semiaxis $0 \le x < \infty$. Consider now the Schrödinger equation for this system:

$$H_\psi = E\psi ; \tag{4.47}$$

where $H = -d^2/dx^2 + V(x)$; $\psi(0) = 0$.

We introduce functions belonging to the continuous spectrum and satisfying (4.47) and the following asymptotic condition:

$$\psi x \overset{(x,k)}{\to} \infty = \frac{A(k)}{k} \sin[kx + \delta(k)] , \quad k = \sqrt{E} , \tag{4.48}$$

as well as the condition $\psi'(0, k) = 1$.

Further, we shall need a function that may be called the S-matrix for the motion under consideration, since $|S - 1|^2$ determines the probability of the process:

$$S(k) = 1 + \int_0^\infty \exp(ikx)V(x)\psi(x, k)dx = A(k)\exp[i\delta(k)] . \tag{4.49}$$

As k approaches infinity, the following estimate for $S(k)$ is obtained:

$$S(k) = 1 - \frac{1}{2ik} \int_0^\infty V(x)dx + \frac{V(0)}{4k^2} - \frac{1}{8k^2} \left(\int_0^\infty V(x)dx\right)^2 + O\left(\frac{1}{k^3}\right) . \tag{4.50}$$

(This can be obtained, for instance, by integrating the integral equation corresponding to (4.47, 48) for the function $\psi(x, k)$ as k tends to infinity.) Hence we obtain the estimates for the phase $\delta(k)$ and amplitude $A(k)$:

$$\delta(k) = \frac{1}{2k} \int_0^\infty V(x)dx + O\left(\frac{1}{k^3}\right)$$

$$\ln A = \frac{1}{4k^2} V(0) + O\left(\frac{1}{k^3}\right) . \qquad (4.51)$$

To derive the trace formulae the following representation will be needed:

$$S(k) = \exp\left[\frac{1}{\pi} \int_0^\infty \frac{\delta(\sqrt{Z})dZ}{Z - k^2}\right] \prod_{l=1}^n \ln \frac{k^2 + \kappa_l^2}{k^2} , \qquad \mathrm{Im}\, k > 0 , \qquad (4.52)$$

where κ_l^2 is the eigenvalue of the Hamiltonian (4.47). This representation can be obtained by using the analytic properties of $S(k)$ in the complex energy plane.

We define the trace of the operator H as

$$\mathrm{Sp} H = \int \langle x|H|x\rangle dx . \qquad (4.53)$$

Using the spectral expansion of the Hamiltonian H

$$H = -\sum_{l=1}^n \kappa_l^2 |l\rangle\langle l| + \int_0^\infty E|\psi_E\rangle\langle\psi_E| dE , \qquad (4.54)$$

where $\langle x|l\rangle$ is an eigenvalue of the discrete spectrum, and $\langle x|\psi_E\rangle \equiv \psi(x,k)$ is a function belonging to the continuous spectrum, one can, by direct calculation, obtain the expression

$$\mathrm{Sp}(H_1 - H_2) = -\sum_{l=1}^{n_1} \kappa_l^{(1)^2} + \sum_{l=1}^{n_2} \kappa_l^{(2)^2} + \frac{2}{\pi} \int_0^\infty k[\delta_1(k) - \delta_2(k)]dk \qquad (4.55)$$

under the condition that $\int_0^\infty [V_1(x) - V_2(x)]dx = 0$. Taking this condition into account the estimate for the phase (4.51) we obtain

$$\int_0^\infty E[\delta_1(E) - \delta_2(E)]dE = -\lim_{k\to\infty} k^2 P \int_0^\infty \frac{E[\delta_1(E) - \delta_2(E)]}{E^2 - k^2}dE . \qquad (4.56)$$

Taking the logarithm of (4.52), we obtain from (4.56):

$$-\sum_{l=1}^{n_1} \kappa_l^{(1)^2} + \sum_{l=1}^{n_2} \kappa_l^{(2)^2} + \frac{2}{\pi} \int_0^\infty E[\delta_1(E) - \delta_2(E)]dE$$

$$= \lim_{k\to\infty} k^2 \left[\ln \frac{A_2(k)}{A_1(k)}\right] . \qquad (4.57)$$

Finally, substituting into the right-hand side of (4.57) the estimate (4.51) for the amplitude A we get the desired sum rule:

$$-\sum_{i=1}^{n_1} \kappa_l^{(1)^2} + \sum_{l=1}^{n_2} \kappa_l^{(2)^2} + \frac{2}{\pi} \int_0^\infty k[\delta_1(k) - \delta_2(k)]dk = \frac{1}{4}[V_2(0) - V_1(0)]. \qquad (4.58)$$

We now proceed to deduce the sum rules in the three-dimensional case of the two-body problem. We introduce the Hamiltonian of the system $H = H_0 + V$. Let there be n levels E_l in the potential V. We define the spectral shift function

$$\Gamma(Z) \equiv \mathrm{Sp}[G(Z) - G_0(Z)] \,, \tag{4.59}$$

where

$$G(Z) = (H - Z)^{-1}; \; G_0(Z) = (H_0 - Z)^{-1} \,.$$

By using the spectral expansions for the Green functions $G(E)$ and $G_0(E)$ and the asymptotic behavior of the functions belonging to the continuous spectrum $\psi(r, Z)$ as r tends to infinity, for the spectral-shift function at $\mathrm{Im}\, Z \neq 0$ one may obtain

$$\lim_{\varepsilon \downarrow 0} \mathrm{Im}\, \mathrm{Sp}[G(Z + i\varepsilon) - G_0(Z + i\varepsilon)] = \frac{1}{2i} \frac{d}{dZ} \mathrm{Sp}\, \ln E(Z), Z > 0 \,, \tag{4.60}$$

where m_l are the normalizing constants of the wave functions belonging to the discrete spectrum. Considering the integral equation for the two-particle Green function, one can write the spectral-shift function

$$\mathrm{Sp}[G(Z) - G_0(Z)] = \frac{1}{8\pi i \sqrt{Z}} \left[\int dy V(y) - \int dy\, dz \exp\left(i\sqrt{Z}|y - z|\right) \right.$$
$$\times \quad \left. V(y)V(z)G(y, z, Z) \right] \,. \tag{4.61}$$

and hence obtain directly the high-energy ($|Z| \to \infty$) behavior of the spectral-shift function

$$\mathrm{Sp}[G(Z) - G_0(Z)] \approx \frac{1}{8\pi i \sqrt{Z}} \int dy V(y)$$
$$\times \quad \frac{1}{2i Z^{3/2}} \sum_{l=0}^{\infty} \frac{(2l + 1)Q_l}{Z^l} \,, \tag{4.62}$$

where the expansion coefficients Q_l are determined through the potential $V(r)$ by the following recursion formulas:

$$Q_l = \frac{1}{2l + 1} \left(\frac{-1}{4}\right)^{l+1} \frac{1}{4\pi} \int dx\, \Omega_{2l+3}^{(1)}(x, x) \,;$$

$$\Omega_n^{(m+1)}(x, x') = 2(m + 1)\Omega_n^{(m+2)}(x, x') + \frac{1}{2} \int_{-1}^{1} d\eta \left(\frac{1 - \eta}{a}\right)^m$$
$$\times \left\{ \Delta_y \Omega_n^{(m)}(y, x') - V(y)\Omega_n^{(m)}(y, x') - m(m + 1)\Omega_n^{(m+2)}(y, x') \right\} \,;$$
$$n = 0, 1, \ldots \,, \quad m = 0, 1, \ldots, n \,, \tag{4.63}$$

where

$$y = \frac{(1-\eta)}{2}x + \frac{(1+\eta)}{2}x' \; ; \quad \Omega_0^{(0)} = 1 \; ;$$

$$\Omega_{n+1}^{(0)} = \Omega_{2s}^{(2t+1)} = \Omega_{2s-1}^{(2t)} = 0 \; ; \quad s = 1,2,\dots \; ; \quad t = 0,1,\dots \; . \tag{4.64}$$

From (4.60) and the asymptotic estimate of the spectral-shift function (4.62) it follows that for the function $f(Z)$,

$$f(Z) \equiv \mathrm{Sp}[G(Z) - G_0(Z)] - \frac{1}{8\pi i\sqrt{Z}} \int d\boldsymbol{x} V(\boldsymbol{x}) \tag{4.65}$$

the following relation holds:

$$\frac{1}{2\pi i} \int_{C_R} Z^s f(Z)dZ = -\sum_{l=1}^{n} m_l E_l^s \; , \tag{4.66}$$

where the contour C_R in the complex Z plane has the form depicted in Fig. 4.1, under the condition that $\mathrm{Re}\{s\} < \frac{1}{2}$.

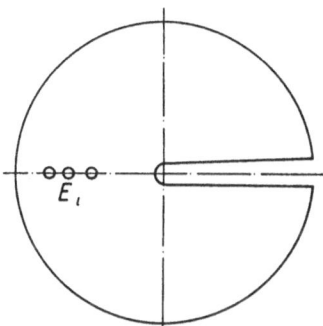

Fig. 4.1. The contour of integration in the left-hand side of (4.66)

Increasing R towards infinity and contracting the part of the contour emcompassing the cut to the semiaxis $[0,\infty)$ we obtain from (4.66) the identity

$$\frac{\exp(i\pi s)\cos\pi s}{\pi} \int_0^\infty dE E^s \mathrm{Im} f(E+i0)$$

$$- \frac{\exp(i\pi s)\sin\pi s}{\pi} \int_0^\infty dE E^s \mathrm{Re} f(E) = -\sum_{l=1}^{n} m_l E_l^s \; . \tag{4.67}$$

For $s = 0$ we obtain, with allowance for (4.60) and (4.66), the summation rules

$$\sum_{l=1}^{n} m_l = -\frac{1}{\pi} \int_0^\infty dE \, \mathrm{Im} f(E+i0) = -\frac{1}{2\pi i} \mathrm{Sp} \ln S(0)$$

$$= -\frac{1}{2\pi i} \ln \det S(0) \; . \tag{4.68}$$

Expression (4.67) permits analytic continuation into the region $s > \frac{1}{2}$. As a result, once again making use of (4.60, 65) for integer s, we obtain the set of summation rules

$$\sum_{l=1}^{n} m_l E_l^{\mu} = \frac{\mu}{2\pi i} \int_0^{\infty} dE E^{\mu-1} \left[\text{Sp} \ln S(E) \right.$$

$$\left. + \frac{2i\sqrt{E}}{4\pi} \int dr V(r) + \frac{2i}{\sqrt{E}} \sum_{l=0}^{\mu-1} \frac{Q_l}{E^l} \right] , \qquad (4.69)$$

$$\mu = 1, 2, \ldots .$$

(Special cases of the sum rules (4.69) are considered in [87].)

For example, for $\mu = 1$ (4.69) becomes

$$\sum_{l=1}^{n} m_l E_l = \frac{1}{2\pi i} \int_0^{\infty} dE \left[\text{Sp} \ln S(E) + \frac{2i\sqrt{E}}{4\pi} \right.$$

$$\left. \times \int dr V(r) + \frac{\int dr V^2(r)}{8\pi i \sqrt{E}} \right] . \qquad (4.70)$$

We shall call (4.69) the *generalized Levinson theorem* [88].

The properties of the spectral-shift function can be utilized to find the *virial coefficients* of a gas of particles interacting by means of a pair potential $V(r)$. We recall that the virial coefficients determine the expansion of the thermodynamic potential Ω or of the pressure P in powers of density. For example, for the pressure, the virial expansion is [89]

$$P = \frac{RT}{V} \left(1 + \frac{R a_2(T)}{V} + \frac{R^2 a_3(T)}{V^2} + \cdots \right) , \qquad (4.71)$$

where a_2 and a_3 represent the second and the third virial coefficients, respectively. Using the technique of the summation rules we shall demonstrate that the second virial coefficient a_2 is expressed through the two-particle scattering phases on the potential $V(r)$ [90, 91]. These results have been generalized for the third virial coefficient in [86, 92].(For a recent review see [P 6].)

Thus, for the second virial coefficient we have

$$a_2(T) = -\sqrt{2}\lambda^2 \text{Sp}[\exp(-\beta h) - \exp(-\beta h_0)] , \qquad (4.72)$$

where T is the temperature; $\lambda = [2\pi\hbar^2/(\mu kT)]^{1/2}$; $h = h_0 + V$ is the two-particle Hamiltonian (k is the Boltzmann constant). (Expression (4.72) can be obtained from the definition for the thermodynamic potential Ω and its virial expansion [89].)

To find the right-hand side of (4.72) we use

$$\exp(-\beta h) = -\frac{1}{2\pi i} \int_C \exp(-\beta Z) G(Z) dZ . \qquad (4.73)$$

Here $G(Z)$ is the two-particle Green function, and the contour C has the form depicted in Fig. 4.2.

Using (4.61, 72, 73) the second virial coefficient is

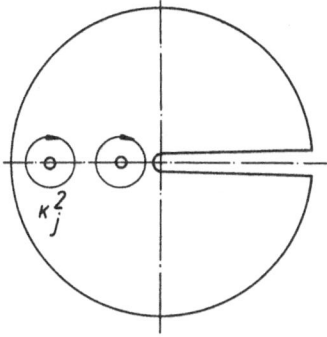

Fig. 4.2. Contour of integration in the right-hand side of (4.73)

$$a_2 = -\sqrt{2}\lambda^3 \left\{ \sum_{j=1}^{N_b} \exp(\beta\kappa_j^2) + \frac{1}{2\pi} \int_0^\infty dE \exp(-\beta E)\, \mathrm{sp}\, q(E) \right\} . \qquad (4.74)$$

In (4.74) the notation $q(E) = -is^+(E)\frac{d}{dE}s(E)$ is introduced, and the operator $s(E)$ is expressed in terms of the S-matrix:

$$\langle p|S|p'\rangle = \frac{\delta(E'-E)}{p\mu}\langle p|s(E)|p'\rangle . \qquad (4.75)$$

Further, for $\mathrm{sp}\, q(E)$ we have

$$\mathrm{sp}\, q(E) = 2\sum_l (2l+1)\frac{d\delta_l(E)}{dE} , \qquad (4.76)$$

where δ_l is the scattering phase. From (4.76) it follows that the trace sp is taken over the angles, contrary to the trace Sp which is taken over the entire volume.

The operator $q(E)$ is called the operator of time delay. Indeed, if one expresses the difference Δt_l between the time a particle having velocity v and angular momentum l travels in the region of the short-range interaction potential, and the time it travels through the same region in the absence of a potential

$$\Delta t_l = \lim_{R\to\infty} \int^R \left[\psi_l^*(r)\frac{r}{v}\psi_l(r) - \frac{2}{4\pi v r^2} \right] dr , \qquad (4.77)$$

then it turns out that [93]

$$\Delta t_l = -s_l^+ ds_l/dE = q_l(E) , \qquad (4.78)$$

i.e., the partial harmonic of the operator $q(E)$ coincides with the time that can be called the delay time, or the collision time. Combination of (4.74,76) leads to an expression for the second virial coefficient in terms of the characteristics of higher partial waves of the two-particle problem, κ_j and $\delta(E)$ [91].

4.3 Systems Involving Variable Numbers of Particles

Up until now we have dealt with systems in which the number of particles in the scattering process remains constant. Clearly such description is approximate. For instance, in the case of elastic scattering of pions on a nucleus there may be intermediate mesonless states or states with two mesons, and so on, present. To take into account the possibility of such transitions, it is necessary to adopt the field-theoretical description of the scattering process. (A detailed discussion of the problems of pion-nucleus interaction may be found in [94,95].) Such an approach permits us, at least in principle, to eliminate the so-called effects of double counting in the potential description of pion-nucleus interaction. What are these effects? Consider, for example, πd-scattering, which we shall describe within the framework of the Faddeev equations giving phenomenological πN- and NN-potentials. It is clear that the diagram presented in Fig. 4.3 contributing to the πN-scattering, after iteration will give an additional contribution to the NN-interaction (Fig. 4.4). Now, since the phenomenological πN- and NN-potentials are chosen independently, it happens that one diagram is included twice.

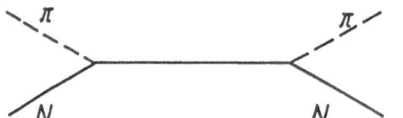

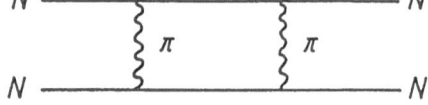

Fig. 4.3. Diagram of πN-interaction   Fig. 4.4. One-meson diagram of NN-interaction

Attempts for an approximate field-theoretical description of the interaction of pions with nuclear systems were made in two directions. In a series of papers a set of two equations was considered explicitly for the case of purely nucleon and nucleon-one-meson components of the wave function of the $\pi 2N$ system [96]. The authors succeded in finding the real and imaginary parts of the πd-scattering lengths. It has not been possible until now to extend this formulation to more complicated systems for two reasons. The first is due to the difficulties in finding the "potential" part of the scattering amplitude. For example, in the case of $\pi 3N$ system this part of the amplitude satisfies the set of Yakubovsky equations. The second reason is related to the absence, in this approach, of a clear understanding of how one must carry out renormalization of the nucleon masses as the number of nucleons increases.

Another field-theoretical formulation of the interaction of pions with nuclei is based on the Low representation for the scattering amplitude. As in the case of the πN-system, this representation leads to equations of the Chew–Low type for the pion-nucleus scattering amplitudes, if conditions similar to the two-particle unitarity for the πN-system are imposed. The clearest formulation of this approach is given in [97], which we follow.

Thus, let the Hamiltonian H describe a boson field and the target. We shall assume that the interacting boson field and the target are described by vector states, which are eigenstates of H:

$$H|k\rangle = \omega_k |k\rangle , \tag{4.79}$$

$$H|\varphi_\alpha\rangle = E_\alpha |\varphi_\alpha\rangle , \tag{4.80}$$

where ω_k is the energy of the boson, and E_α and $|\varphi_\alpha\rangle$ are, respectively, the energy and eigenvector of the target, which is in the state α. Consider an eigenvector $|\psi_{k|\alpha}\rangle$ describing a boson interacting with the target, the boson of momentum k being incident on the target in the state α. This eigenvector satisfies the Schrödinger equation

$$(E_\alpha + \omega_k - H)|\psi_{k,\alpha}\rangle^\pm = 0 . \tag{4.81}$$

The signs $\pm$ correspond to the outgoing and the incoming waves. The total energy E equals $E = E_\alpha + \omega_k$. Now we define the boson creation and absorption operators b^+ and b:

$$b^+(k)|0\rangle \equiv |k\rangle , \tag{4.82}$$

i.e., the operator $b^+(k)$ acting on vacuum produces a boson in the state $|k\rangle$. Taking into account the boundary condition at $-\infty$ we represent the vector $|\psi_{k,\alpha}\rangle$ as

$$|\psi_{k,\alpha}\rangle^\pm = b^+(k)|\varphi_\alpha\rangle + |\chi\rangle^\pm . \tag{4.83}$$

Clearly, $|\chi\rangle^\pm$ corresponds to the scattered wave. Substituting (4.83) into (4.81) we obtain for $|\chi\rangle$

$$|\chi\rangle^\pm = G(E_\alpha + \omega_k \pm i\eta) J(k)|\varphi_\alpha\rangle , \tag{4.84}$$

where

$$G(Z) \equiv (Z - H)^{-1} \tag{4.85}$$

is the Green function of the system, and

$$J(k) \equiv [H, b^+(k)] - \omega_k b^+(k) . \tag{4.86}$$

Thus, the eigenvector of H describing the scattering of a boson of momentum k incident on the target is

$$|\psi_{k,\alpha}\rangle^\pm = b^+(k)|\varphi_\alpha\rangle + G(E_\alpha + \omega_k \pm i\eta) J(k)|\varphi_\alpha\rangle . \tag{4.87}$$

Knowing the state vectors $|\psi_{k,\alpha}\rangle^\pm$ we find the S-matrix of the process on the basis of the definition

$$\langle k', \beta | S | k, \alpha \rangle \equiv^- \langle \psi_{k,\beta} | \psi_{k',\alpha} \rangle^+ . \tag{4.88}$$

To this end we take advantage of the relation between the Green functions, the validity of which is easily verified:

$$G(x + i\eta) = G(x - i\eta) - 2\pi i \delta(x - H) . \tag{4.89}$$

Hence, we obtain the relationship between $|\psi\rangle^+$ and $|\psi\rangle^-$:

$$|\psi_{k,\alpha}\rangle^+ = |\psi_{k,\alpha}\rangle^- - 2\pi i \delta(E_\alpha + \omega_k - H)J(k)|\varphi_\alpha\rangle . \tag{4.90}$$

Substituting (4.90) into definition (4.88) of the S-matrix we obtain

$$\begin{aligned}
\langle k', \beta|S|k, \beta\rangle &= {}^-\rangle\psi_{k',\beta}|\psi_{k,\alpha}\rangle^- - 2\pi i^- \langle \psi_{k',\beta}|\delta(E_\alpha + \omega_k + H)J(k)|\varphi_\alpha\rangle \\
&= \delta_{\alpha\beta}\delta(k' - k) - 2\pi i \,\delta(E_\alpha + \omega_k \\
&\quad - E_\beta - \omega_{k'})\langle k', \beta|T(E_\beta + \omega_{k'})|k, \alpha\rangle .
\end{aligned} \tag{4.91}$$

Equation (4.91) may be considered as the definition of the T-matrix for systems with a variable number of particles, i.e.,

$$\langle k', \beta|T(E_\beta + \omega_{k'})|k, \alpha\rangle \equiv{}^- \psi_{k',\beta}|J(k)|\varphi_\alpha\rangle . \tag{4.92}$$

Note that (4.92) determines the half-mass-shell matrix T. Substitution of the vector $|\psi\rangle^-$ into (4.88) leads to the second half-mass-shell T-matrix:

$$\langle k', \beta|T(E_\alpha + \omega_k|k, \alpha\rangle \equiv \langle \varphi_\beta|J^+(k')|\psi_{k,\alpha}\rangle^+ . \tag{4.93}$$

Obviously, both definitions (4.92, 93) coincide on the mass shell, i.e., when $E_\alpha + \omega_k = E_\beta + \omega_{k'}$.

To obtain the Low representation for the T-matrix, we substitute the eigenvector $|\psi_{k,\alpha}\rangle$ from (4.87) into (4.93):

$$\begin{aligned}
\langle k', \beta|T(E_\alpha + \omega_k)|k, \alpha\rangle &= \langle \varphi_\beta|J^+(k')b^+(k)|\varphi_\alpha\rangle \\
&\quad + \langle \varphi_\beta|J^+(k')G(E_\alpha + \omega_k + i\eta)J(k)|\varphi_\alpha\rangle .
\end{aligned} \tag{4.94}$$

Using the identity

$$J^+(k')b^+(k) \equiv b^+(k)J^+(k') + [J^+(k'), b^+(k)]_- , \tag{4.95}$$

we rewrite (4.94) in the form

$$\begin{aligned}
\langle k', \beta|T(E_\alpha + \omega_k)|k, \alpha\rangle &= \langle \varphi_\beta|b^+(k)J^+(k')|\varphi_\alpha\rangle \\
&\quad + \langle \varphi_\beta|[J^+(k'), b^+(k)]_-|\varphi_\alpha\rangle \\
&\quad + \langle \varphi_\beta|J^+(k')G(E_\alpha + \omega_k + i\eta)J(k)|\varphi_\alpha\rangle .
\end{aligned} \tag{4.96}$$

We transform the first term in the right-hand side of (4.96) by using

$$J^+(k)|\varphi_\beta\rangle = \{E_\beta - \omega_k - H\}b(k)|\varphi_\beta\rangle \tag{4.97}$$

or

$$b(k)|\varphi_\beta\rangle = [E_\beta - \omega_k - H - i\eta]^{-1}J^+(k)|\varphi_\beta\rangle . \tag{4.98}$$

[By a direct calculation one may verify that transition from (4.97) to (4.98) is possible only for theories in which meson states with negative energy are absent.] Thus, the first term in (4.96) becomes

$$\langle \varphi_\beta | b^+(k) J^+(k') | \varphi_\alpha \rangle = \langle \varphi_\beta | J(k) G(E_\beta - \omega_k - i\eta) J^+(k') | \varphi_\alpha \rangle \ . \tag{4.99}$$

In the theory of πN-scattering such terms are usually known as crossing-terms. Using (4.99) we obtain the desired Low representation for the half-mass-shell T-matrix:

$$
\begin{aligned}
\langle k', \beta | T(E_\alpha + \omega_k) | k, \alpha \rangle = &\langle \varphi_\beta | J(k) G(E_\beta - \omega_k - i\eta) J^+(k') | \varphi_\alpha \rangle \\
&+ \langle \varphi_\beta | J^+(k') G(E_\alpha + \omega_k + i\eta) J(k) | \varphi_\alpha \rangle \\
&+ \langle \varphi_\beta | J^+(k'), b^+(k)]_- | \varphi_\alpha \rangle \ .
\end{aligned}
\tag{4.100}
$$

Clearly, if in the right-hand side of (4.100) the completeness relation is utilized in the form

$$1 = \sum_\gamma |\varphi_\gamma\rangle\langle\varphi_\gamma| + \sum_{\gamma'} \int dq |\psi_{q,\gamma}\rangle\langle\psi_{q,\gamma}| \ , \tag{4.101}$$

i.e., only two-particle unitarity is taken into account, we obtain nonlinear equations for the transition amplitudes T.

Upon performing simple transformations, taking into account only the commutation properties of the operators $b(k)$ and $J(k')$, one can arrive at the Low representation for the completely off-mass-shell T-matrix:

$$
\begin{aligned}
\langle k', \beta | T(E) | k, \alpha \rangle = &\langle \varphi_\beta | J^+(k') G(E - i\eta) J(k) | \varphi_\alpha \rangle \\
&+ \langle \varphi_\beta | J(k) G(E_\alpha + E_\beta - E - i\eta) J^+(k') | \varphi_\alpha \rangle \\
&+ \langle \varphi_\beta | [b(k'), J(k)]_- | \varphi_\alpha \rangle \ .
\end{aligned}
\tag{4.102}
$$

Now let us see what becomes of the Low representation (4.102) for the T-matrix, if one goes into details of the total Hamiltonian for various models of the πN-interaction.

So, let the pion-nuclear Hamiltonian be

$$H = H_\pi + H_N + h_{\pi N} \ , \tag{4.103}$$

where

$$H_\pi = \sum_k \omega_k b^+(k) b(k) \tag{4.104}$$

is the free pion Hamiltonian ($\omega_k = \sqrt{k^2 + m_\pi^2}$); H_N is the nucleon Hamiltonian including all nucleon interactions except interaction with pions. Making use of this property of H_N we obtain

$$[H_N, b(k)]_- = [H_N, b^+(k)]_- = 0 \ , \tag{4.105}$$

which immediately leads to

$$J(k) \equiv [H, b^+(k)]_- - \omega_k b^+(k) = [h_{\pi N}, b^+(k)]_- \ . \tag{4.106}$$

Now as $h_{\pi N}$ let us choose a simple Hamiltonian linear in the meson field:

$$h_{\pi N} = \sum_{k} [V(k)b(k) + V^+(k)b^+(k)] , \qquad (4.107)$$

where the operator V depends only on nucleon operators. [The Hamiltonian (4.107) can be easily reduced to the Chew–Low Hamiltonian by assigning an isotopic index to the pion operators $b(k)$ and by the appropriate choice of operators $V(k)$.] The Hamiltonian (4.107) leads to

$$J(k) = [h_{\pi N}, b^+(k)]_- = V(k) , \qquad (4.108)$$

from which it follows that the last term in the Low representation (4.102) vanishes and it becomes

$$\langle k', \beta | T(E) | k, \alpha \rangle = \langle \varphi_\beta | V^+(k')G(E + i\eta)V(k) | \varphi_\alpha \rangle$$
$$+ \langle \varphi_\beta | V(k)G(E_\alpha + E_\beta - E - i\eta)V^+(k') | \varphi_\alpha \rangle . \qquad (4.109)$$

It is easy to check that the condition of crossing symmetry, which in the nonrelativistic case is

$$\langle k', \beta | T(E) | k, \alpha \rangle = \langle -k', \alpha | T(E_\alpha + E_\beta - E) | - k, \beta \rangle^*$$
$$= \langle -k, \beta | T^+(E_\alpha + E_\beta - E) | - k', \alpha \rangle , \qquad (4.110)$$

leads to the following restriction being imposed on the form of the operators $V(k)$:

$$V(k) = \exp(i\varphi)V^+(-k) . \qquad (4.111)$$

For Hamiltonians of the Chew–Low type the relation $V(k) = V^+(-k)$ is usually true.

Now let us consider a "quadratic" in the pion field πN Hamiltonian of the "potential" type:

$$h_{\pi N} = \sum_{qq'} W(q, q')b^+(q)b(q') . \qquad (4.112)$$

With the Hamiltonian (4.112) we obtain for the function $J(k)$

$$J(k) = \sum_{q} W(q, k)b^+(q) . \qquad (4.113)$$

Using (4.113) for the commutator in the third term of the Low representation (4.102) we obtain

$$[b(k'), J(k)]_- = W(k', k) , \qquad (4.114)$$

while the representation itself assumes the form

$$\langle \mathbf{k}', \beta | T(E) | \mathbf{k}, \alpha \rangle = \langle \varphi_\beta | W(\mathbf{k}', \mathbf{k}) | \varphi_\alpha \rangle$$
$$+ \sum_{q'q} \langle \varphi_\beta | b(\mathbf{q}') W(\mathbf{q}', \mathbf{k}') G(E - i\eta) W(\mathbf{q}, \mathbf{k}) b^+(\mathbf{q}) | \varphi_\alpha \rangle$$
$$+ \sum_{q'q} \langle \varphi_\beta | b^+(\mathbf{q}) W(\mathbf{q}, \mathbf{k}) G(E_\alpha + E_\beta - E + i\eta)$$
$$\times W(\mathbf{q}', \mathbf{k}') b(\mathbf{q}') | \varphi_\alpha \rangle . \tag{4.115}$$

The first two terms in (4.115) are characteristic of theories of the potential type; for instance, the first term can be interpreted as the Born term of potential scattering. The third summand of the crossing type in (4.115) is usually absent in the theory of potential interaction of pions with nuclei. It is quite clear that its absence is related to the additional hypothesis of the absence of particles in the target identical with the incident one, i.e.,

$$b(\mathbf{k}') | \varphi_\alpha \rangle = 0 . \tag{4.116}$$

Thus, when describing the interaction of pions with nuclei within the framework of potential theory, one always deals with the implicit existence of the hypothesis that no pions are present in the nucleus.

5. Outlook

As mentioned in the preface, the proper place and importance of the few-body problem can only be appreciated by considering it within the context of the respective many-body problem [98]. Of course, in such a concise treatment as the present one it is impossible to do justice to all aspects of the few-body problem. For further reading and acquaintance with related formal structures and techniques [P 2].

In addition it seems appropriate to mention those more recent investigations that have not been dealt with in these lectures. First of all, this concerns the work on application of the method of hyperspherical functions (K-harmonics) to the studies of properties of the lightest nuclei [99]. This method has been shown to be extremely efficient for calculations involving central short-range potentials without strong repulsion at short distances. More complicated potentials necessitate modification of the method. Recently, the matrix generalization of this approach has been proposed which is applied in field-theoretical models [100].

Very beautiful results were obtained within the new formulation of the boundary condition model in the three-body problem [101]. It has been shown that to obtain the Faddeev equations in this case the information given by the conditions of the Dirichlet problem is insufficient, contrary to what has earlier been assumed, but it is necessary to add information on the internal Neumann problem.

Significant progress in calculations of properties of few-particle systems, including those with purely Coulomb interaction (such, for example, as μ-mesic molecules), has been recently achieved by application of various modifications of the Monte-Carlo method [22, 102].

Finally, we mention the application of the method of path integration in the theory of few-body systems [103]. The principal idea of using this method is that when the path integral is made discrete, the number of points at which the corresponding functions are computed increases with the increase of the number of particles (i.e., the number of dynamic variables) much slower than does the number of points when the initial dynamic equations themselves are directly rendered discrete.

General Reading

P1 L.D. Faddeev, S. Merkuryev: *Quantum Scattering Theory for Few-Particle Systems* (Nauka, Moscow 1985) [in Russian]

P2 W. Glöckle: *The Quantum Mechanical Few-Body Problem.* Texts and Monogr. Physics (Springer, Berlin, Heidelberg 1983)

P3 T.K. Lim, C.G. Bao, D.P. Hou, H.S. Huber (eds.): *Few Body Methods. Principles and Applications* (World Scientific, Singapore 1986)

P4 R.G. Newton: *Scattering Theory of Waves and Particles.* Texts and Monogr. Physics 2nd Ed. (Springer, New York 1982)

P5 E. Schmid, H. Zigelman: *The Quantum Mechanical Three-Body Problem* (Vieweg, Braunschweig 1974)

P6 A.G. Sitenko: *Scattering Theory* (Springer, Berlin, Heidelberg 1990)

References

1   L.D. Faddeev: in *Collected Papers of the Institute of Mathematics of the USSR Academy of Science* 1963, p. 69–119 [in Russian].
      L.D. Faddeev: JETP **12**, 1014 (1960); Dokl. Akad. Nauk **6**, 384 (1961); ibid **7**, 600 (1962); *Mathematical Aspects of the Three-Body Problem in the Quantum Scattering Theory* (Isr. Prog. Sci. Trans, Jerusalem 1965)
2   H.P. Noyes: Phs. Rev. Lett. **15**, 538–540 (1965)
3   O.O. Kowalski: Phys. Rev. Lett. **15**, 798-800 (1965)
4   T. Osborn: Nucl. Phys. A **138**, 305 (1969)
5   M.K. Srivastava, D.W.L. Sprung: *Advances in Nuclear Physics*, Vol. 8, ed. by M. Baranger, E. Vogt (Plenum, New York 1975), p. 121. W. Plessas: in [P3], p. 43
6   B. Akhmadkhodzhaev, V.B. Belyaev, J. Wrzechionko: JETP Lett. **9**, 692–694 (1969) [in Russian]; V.B. Belyaev, A.L. Zubarev: Yad. Fiz. **14**, 545–558 (1971) [in Russian]; V.B. Belyaev, J. Wrzechionko, A.L. Zubarev: JINR Dubna Preprint E4–5763 (1971)
7   Y. Yamagushi: Phys. Rev. **95**, 1628–1634 (1954)
8   V.B. Belyaev, J. Wrzechionko, B.F. Irgaziev: Yad. Fiz. **20**, 1267–1272 (1974); ibid **24**, 1250–1255 (1976) [in Russian]
9   A.L. Zubarev: *The Schwinger Variational Principle in Quantum Mechanics* (Energoizdat, Moscow 1981) [in Russian]
10   V.N. Efimov: JINR Dubna Preprint 4–5741 (1971); JINR Dubna Preprint P-2546 (1966) [in Russian]
11   Yu.V. Vorob'ev: *The Method of Moments in Applied Mathematics* (GIFML, Moscow 1958) [in Russian]
12   T.A. Osborn: Nucl. Phys. **A211**, 211–220 (1973)
13   S. Weinberg: Phys. Rev. **131**, 440–462 (1963); R. Jost, A. Pais: Phys. Rev. **82**, 840–855 (1951)
14   E. Harms: Phys. Rev. **C1**, 1667–1681 (1970)
15   L.D. Faddeev, S.P. Merkuriev: *Quantum Scattering Theory for Few-Particle Systems* (Nauka, Moscow 1985) [in Russian]
16   A. Sommerfeld: *Mechanik*, (Akad. Verl., Leipzig 1944)
17   J.D. Walecka, L.L. Foldy: Ann. Phys. **54**, 447–468 (1969)
18   H. Poincaré: *Les Méthodes nouvelles de la Mécanique Celeste*, Vol. 1 (Gauthier-Villars, Paris 1892). W. Aurel: *The Analytical Foundations of Celestial Mechanics*, (Princeton University Press, Princeton 1941)
19   L. Thomas: Phys. Rev. **47**, 903–915 (1935)
20   R.A. Minlos, L.D. Faddeev: JETP **41**, 1850–1855 (1961) [in Russian]
21   V. Efimov: Phys. Lett. **33B**, 563–564 (1970); Nucl. Phys. **A210**, 157 (1973);
22   V.I. Kukulin, V.G. Neudachin, V.N. Pomerantsev: Yad. Fiz. **24**, 298–307 (1976) [in Russian]; J. Phys. G Nucl. Phys. **4**, 1709–1723 (1978) L.S. Ferreira, A.C. Fonseca, L. Streit (eds.): *Models and Methods in Few-Body Physics*, Lect. Notes Phys., Vol. 273 (Springer, Berlin, Heidelberg 1987)
23   J.L. Ballot, M. Fabre de la Ripelle (eds.): *Few-Body Problems in Particle, Nuclear, Atomic, and Molecular Physics* (Few-Body Systems, Suppl. 1) (Springer, Berlin, Heidelberg 1986)
24   S.P. Merkuriev: Yad. Fiz. **24**, 289–301 (1976) [in Russian]
25   S.P. Merkuriev, A.K. Motovilov: in *Theory of Quantum Systems with Strong Interactions* (Kalinin University Press, Kalinin 1983), p. 95–103 [in Russian]

26   L.D. Faddeev: *Lectures at the School of the Moscow Engineering Physics Institute* (MEPhI, Moscow 1971) [in Russian]

27   R.D. Amado, J.V. Noble: Phys. Lett. **835**, 25 (1971); Phys. Rev. **D5**, 1992-2011 (1972)

28   Yu.A. Simonov, I.L. Grach, M.Zh. Shmatikov: Nucl. Phys. **A334**, 82–98 (1980)

29   H.P. Noyes, H. Fideldey: in *Three Particle Scattering in Quantum Mechanics*, ed. by J. Gillespie, J. Nutall (Benjamin, New York 1968), p. 195; J.S. Levinger, A.H. Lu, R. Stagat: Phys. Rev. **179**, 926 (1969)

30   V.B. Belyaev, V.N. Efimov, E.G. Tkachenko, H. Schulz: Yad. Fiz. **18**, 779–791 (1973) [in Russian]

31   G.V. Skornyakov, K.A. Ter–Marterosyan: JETP **31**, 775–792 (1950) [in Russian]

32   V.B. Belyaev, V.N. Efimov: Preprint Institute of Theoretical Physics ITPh-71-78P (Kiev 1971) [in Russian]

33   H. Ziegelman, G. Sohre: Phys. Lett. **34B**, 579–582 (1971)

34   (a) E.O. Alt, P. Grassberger, W. Sandhas: Nucl. Phys. **B2**, 167–183 (1967); (b) A.A. Khelashvili, JINR Dubna Preprint P2–3371 (1967) [in Russian]; (c) W. Sandhas: Acta Phys. Austriaca Suppl. IX, 57 (1972); [P3], p.3

35   C. Lovelace: in *Strong Interactions and High Energy Physics*, ed. by R.G. Moorhouse (Oliver and Boy, London 1964)

36   (a) D.R. Karlsson, E.M. Zeiger: Phys. Rev. **D9**, 1761–1772 (1974); ibid **D10**, 1291–1311 (1974); (b) P. Grassberger, W. Sandhas: Z. Phys. **220**, 29 (1969); (c) T. Osborn, O. Kowalski: Ann. Phys. **68**, 361 (1971)

37   E. Harms: Phys. Lett. **41B**, 26–28 (1972)

38   T.A. Osborn, D. Eyre: Nucl. Phys. **A327**, 125 (1979)

39   C. de Boer: *A Practical Guide to the Splines* (Springer, Berlin, Heidelberg 1978); P.M. Prenter: *Splines and Variational Methods* (Wiley, New York 1975); G.L. Payne: in [21], p. 64; M.H. Shultz: *Spline Analysis* (Prentice-Hall 1973)

40   I.H. Sloan: Phys. Rev. **165**, 1587–1601 (1968)

41   H.M. Nussenzweig: *Causality and Dispersion Relations* (Academic, New York 1972); S. Mandelstam: Phys. Rev. **140**, 375 (1965); M. Rubin, R. Sugar, G. Tiktopoulos: Phys. Rev. **146**, 1130 (1966); ibid. **159**, 1348 (1967); ibid **162**, 1555 (1967); V. Avishai, W. Ebenhöh, A.S. Rinat: Ann. Phys. **55**, 341 (1969); A.S. Rinat, M. Stingl: Ann. Phys. **65**, 141 (1971)

42   G. Barton, A.C. Phillips: Nucl. Phys. **A132**, 97–111 (1969)

43   A.M. Badalyan, Yu.A. Simonov: Yad. Fiz. **21**, 890–903 (1975) [in Russian]

44   N.I. Muskhelishvili: *Singular Integral Equations* (Nauka, Moscow 1968) [in Russian]; F.D. Gakhov, *Boundary Value Problems* (Nauka, Moscow 1977) [in Russian]

45   H. Pagels: Phys. Rev. **B140**, 1599–1613 (1965)

46   J.M. Greben, Yu.A. Simonov: Phys. Rev. **C18**, 642–654 (1978)

47   A.M. Badalyan, Yu.A. Simonov: Yad. Fiz. **21**, 890–903 (1975) [in Russian]

48   D.A. Kirzhnits, D.Yu. Kryuchkov, N.J. Takibaev: in *Elementary Particles and the Atomic Nucleus*, Vol. 10, 741–764 (1979) [in Russian]; V.B. Belyaev, D.A. Kirzhnitz, N.Zh. Takibaev, M.Kh. Khankhasayev: Sov. J. Nucl. Phys. **32**, 578 (1980); V.B. Belyaev, M.Kh. Khankhasayev: Phys. Lett. B **137**, 299 (1984); M.Kh. Khankhasayev: Nucl. Phys. A. **505** 717, (1989)

49   N.J. Takibaev, Preprint IHEP 40–76, Acad. Sci. Kazakh SSR, Alma-Ata, 1976 [in Russian]

50   V.B. Belyaev, O.P. Solovtsova: Yad. Fiz. **33**, 699–711 (1981) [in Russian]

51   V.B. Belyaev, O.P. Solovtsova: Yad. Fiz. **34**, 339–351 (1981) [in Russian]

52   D.A. Kirzhnits: in *Problems in Theoretical Physics. In Memory of I.E. Tamm* (Nauka, Moscow 1972), p. 74–92 [in Russian]

53   V.B. Belyaev, O.P. Solovtsova: J. Phys. G Nucl. Phys. **8**, 349–355 (1982)

54   H.P. Noyes: Phys. Rev. **D5**, 1547–1551 (1972)

55   V.B. Belyaev, S.E. Brener, R.M. Galymianov, A.L. Zubarev: JINR Dubna Preprint P4-82-810, 1982 [in Russian]

56   J.L. Friar, B.F. Gibson, G.L. Payne: Z. Phys. **A301**, 309 (1981); for recent reviews see A.C. Fonseca, in [P6], p.111; in [21], p. 161

57   H. Fiedeldey: in *Few-Body Systems in Particle and Nuclear Physics*, ed. by T. Sasakawa (North-Holland, Amsterdam 1987) p. 335c

58   S.I. Vinitsky et al: JINR Dubna Preprint P4-13-036, 1980 [in Russian]

59   O.A. Yakubovsky: Yad. Fiz. **5**, 1312–1320 (1967) [in Russian]

60   P. Grassberger, W. Sandhas: Nucl. Phys. **B2**, 181 (1967); E. Alt, P. Grassberger, W. Sandhas: in *Few Particle Problems in the Nuclear Interaction*, ed. by I. Staus (North-Holland, Amsterdam 1972) p. 299; Preprint JINR Dubna E4-6688 (1972); W. Sandhas: in *Few Body Dynamics*, ed. by A.N. Mitra (North-Holland, Amsterdam 1976) p. 540; Czech. J. Phys. **B25**, 251 (1975); E. Alt, P. Grassberger, W. Sandhas: Phys. Rev. **C1**, 85 (1970)

61   S.P. Merkuriev, S.L. Yakovlev: in *Reports of the USSR Academy of Sciences* **262**, 591–594 (1982) [in Russian]

62   A.G. Baryshnikov, L.D. Blokhintsev, I.M. Narodetsky: Yad. Fiz. **25**, 1167–1175 (1977) [in Russian]; R. Perne, W. Sandhas: Phys. Rev. Lett. **39**, 788–791 (1977); H. Kröger, W. Sandhas: Phys. Rev. Lett **40**, 834 (1978) V.F. Kharchenko, V.P. Levashov: Nucl. Phys. **343A**, 249–317 (1980); J.A. Tjon: Nucl. Phys. **353 A**, 47–60 (1981)

63   B.F. Gibson, D.R. Lehman: Phys. Rev. **23C**, 404–421 (1981)

64   A.M. Gorbatov, Yu.N. Krylov: in *Theory of Quantum Systems with Strong Interactions* (Kalinin University Press, Kalinin 1983) p. 34–42 [in Russian]; J.L. Ballot, M. Fabre de la Ripelle, J. Navarro: Phys. Lett. **B 143**, 19 (1984); M. Fabre de la Ripelle: in [21]; S.Y. Larsen in [P3]

65   I.M. Narodetsky: Riv. Nuovo Cimento **4**, 125 (1981); Nucl. Phys. **A221**, 191–211 (1974); J.A. Tjon: Phys. Lett **56B**, 217–220 (1975)

66   V.B. Belyaev, K. Möller: Z. Phys. **A279**, 47–56 (1976)

67   V.B. Belyaev, J. Wrzechionko: Yad. Fiz. **28**, 392–402 (1978) [in Russian]

68   V.B. Belyaev, V.V. Pupyshev: Yad. Fiz. **39**, 594–601 (1984) [in Russian]

69   V.B. Belyaev, J. Wrzechionko, M.I. Sakvarelidze: JINR Dubna Preprint E4-11505 (1978); Phys. Lett **B83**, 19–22 (1979)

70   I.R. Afnan, A.W. Thomas: Phys. Rev. **C10**, 109–122 (1974)

71   K.A. Brueckner: Phys. Rev. **98**, 769–771 (1955)

72   R. Abela, G. Backenstoss, A. Brandao D'Oliviera, M. Izycki, H.O. Meyer, I. Schwanner, L. Tauscher, P. Blüm, W. Fetscher, D. Gotta, H. Koch, H. Poth, L.M. Simons: Phys. Lett. **B68**, 429–432 (1977)

73   V.B. Belyaev, V.V. Pupyshev: Yad. Fiz. **35**, 905–913 (1982) [in Russian]

74   V.F. Sears, F.C. Khanna: Phys. Lett. **56B**, 1–3 (1975)

75   J.D. Seagrave, B.L. Berman, T.W. Phillips: Phys. Lett **91B**, 200–202 (1980)

76   V.B. Belyaev, V.V. Pupyshev: Yad. Fiz. **31**, 1324–1331 (1980) [in Russian]

77   C.K. Majumdar: Phys. Rev. **B1**, 287–297 (1970); J. Math. Phys. **19**, 2187–2193 (1978)

78   J.E. Van Himbergen: Physica **A86**, 93–116 (1977)

79   J. Bethe: Z. Phys. **71**, 205–228 (1931)

80   L.R. Dodd: J. Math. Phys. **11**, 207–213 (1970)

81   M. Gell–Mann, M.L. Goldberger, W.E. Tirring: Phys. Rev. **95**, 1612–1627 (1954)

82   L.D. Faddeev: Dokl. Akad. Nauk **115**, 878–881 (1957) [in Russian]

83   V.S. Buslaev, L.D. Faddeev: Dokl. Akad. Nauk **132**, 13–17 (1960) [in Russian]

84   V.S. Buslaev: Dokl. Akad. Nauk **143**, 1067–1070 (1962) [in Russian]

85   V.S. Buslaev: in *Problems of Mathematical Physics*, ed. by M.Sh. Birman (Leningrad University Press, Leningrad 1966) [in Russian]

86   V.S. Buslaev, S.P. Merkuriev: Teor. Mat. Fiz. **5**, 372–387 (1970) [in Russian]; R. Dashen, Shang-Keng Ma, H.J. Berstein: Phys. Rev. **187**, 345 (1969); R. Dashen, Shang-Keng Ma: J. Math. Phys. **11**, 1136 (1970); ibid **12**, 689 (1971)

87   R. Newton: J. Math. Phys. **18**, 1348–1358; Dreyfus: Helv. Phys. Acta **51**, 321–329 (1978); D.A. Kirzhnits, N.J. Takibaev: Yad. Fiz. **16**, 253–268 (1972) [in Russian]

88   P. Beregi, B.N. Zakhariev, S.A. Niyazgulov: Elementary Particles and the Atomic Nucleus, 1973, Vol. 4, p. 512–533 [in Russian]

89   L.D. Landau, E.M. Lifshits: *Statistical Physics* (Nauka, Moscow 1964) [in Russian]

90   M.S. Beth, R.J. Ulenbeck: Physica **3**, 729–742 (1936); ibid **4**, 915 (1937); L. Gropper: Phys. Rev. **50**, 963 (1936)

91   D. Bolle: Ann. Phys. **121**, 131–146 (1979)

92   D.A. Kirzhnits, N.J. Takibaev: JETP **75**, 785–796 (1978) [in Russian]

93   F.T. Smith: Phys. Rev. **118**, 349–356 (1960); ibid **119**, 2098 (1960)

94   D.S. Koltun, J.M. Eisenberg: *Theory of Meson Interactions with Nuclei* (Wiley-Interscience, New York 1979)

95   T.I. Kopalejshvili: *Problems of Pion Interactions with Nuclei* (Energoatomizdat, Moscow 1984) [in Russian]

96   D.S. Koltun, T. Mizutani: Ann. Phys. **1**, 109–140 (1977); Y. Avishay, T. Mizutani: Nucl. Phys. **A338**, 377–382 (1980); T Afnan: Phys. Rev. **C10**, 109 (1974); A.S. Rinat, R. Starkand: Nucl. Phys. **A397**, 381 (1983)

97   E.R. Siciliano, R.M. Thaler: Ann. Phys. **115**, 191–247 (1978)

98   P. Ring, P. Schuck: *The Nuclear Many-Body Problem* (Springer, Berlin, Heidelberg 1980)

99   M. Fabre de la Ripelle: in *Models und Methods in Few-Body Physics*, Lect. Notes Phys. Vol. 273 (Springer, Berlin, Heidelberg 1987)

100   Yu.A. Simonov: Preprint Inst. of Theoretical and Experimental Physics ITEPh 14 (1983)

101   S.P. Merkuriev, A.K. Motovilov: Lett. Math. Phys. **7**, 497–503 (1983)

102   J.G. Zabolitzky: Nucl. Phys. **A416**, 401 (1984)

103   Y. Alhassid, G. Maddison, K. Laganke, K. Chow, S.E. Koonin: Preprint 3074-829 "Path Integral Monte-Carlo Calculations of ^{4}He and ^{6}Li", Yale University (1984)

Subject Index

Adiabatic basis 79
Asymptotic
– conditions 84
– four-particle wave function 88

Cauchy formula 5
Chew-Low Hamiltonian 124
Convergency
– Bateman expansion 20
– Born series 24
– Bubnov–Galerkin expansion 21
– moments method expansion 23, 48
– partial wave expansion 15
– scattering length expansion 65
– spline functions 48, 49

Effect
– Efimov 28, 61
– Thomas 28, 62
Equation
– Alt–Grassberger-Sandhas (AGS)
  41, 46, 87
– approximate 88, 92, 97, 100, 104
– Faddeev–differential 82
– Faddeev–integral 31, 33, 82, 110
– Lippman–Schwinger 2, 3, 7, 8, 29,
  110
– nonlinear 123
– N/D 56
– Skornyakov–Ter–Martirosyan 37
– Yakubovsky–differential 81, 84
– Yakubovsky–integral 81, 84

Ferromagnet 107
Fourier transform 1, 16, 108

Functions
– Green 3
– of spectral shift 116
– spherical Bessel 3
– spline 47, 48

Green functions 3
– spectral expansion 5

Hilbert identities 3, 4
Hilbert–Schmidt
– eigenfunction 24
– expansion 25

Interaction
– NN 103
– pion–nucleus 66

K-matrix approximation 80

Method
– Bateman 13, 35
– Bubnov–Galerkin 20
– dispersion 54
– ECC 62, 64
– momentum 22, 47
– Monte-Carlo 126
– path integration 126

Non-relativistic quark model 80, 103

Off–shell
– t–matrix 3
– wave function 8

Springer Tracts in Modern Physics

Editors: G. Höhler, E. A. Niekisch

Springer-Verlag
Berlin Heidelberg New York London Paris Tokyo Hong Kong

Research Reports in Physics

J. Eberth, University of Cologne; **R. A. Meyer,** Lawrence
Livermore National Laboratory, Livermore, CA;
K. Sistemich, Kernforschungsanlage Jülich (Eds.)

Nuclear Structure of the Zirconium Region

Proceedings of the International Workshop, Bad Honnef,
Fed. Rep. of Germany, April 24–28, 1988

Organized by the Kernforschungsanlage Jülich, Univer-
sität zu Köln, and Lawrence Livermore National
Laboratory

1988. XII, 424 pp. 231 figs. Softcover DM 141,–
ISBN 3-540-50120-7

M. Lozano, M. I. Gallardo, J. M. Arias, University
of Sevilla (Eds.)

Nuclear Astrophysics

Proceedings of the Third International Summer School,
La Rábida, Huelva, Spain, June 19–July 2, 1988

1989. VIII, 355 pp. 126 figs. Softcover DM 98,–
ISBN 3-540-50751-5

B. N. Zakhariev, Lab. of Theoretical Physics, Dubna;
A. A. Suzko, Luikov Heat and Mass Transfer
Institute, Minsk

Direct and Inverse Problems
Potentials in Quantum Scattering

1990. Approx. 200 pp. 42 figs. Softcover, in prep.
ISBN 3-540-52484-3

Springer-Verlag
Berlin Heidelberg New York London Paris Tokyo Hong Kong

Springer